BECOMING WILD

BECOMING WILD

HOW ANIMAL CULTURES RAISE FAMILIES, CREATE BEAUTY, AND ACHIEVE PEACE

CARL SAFINA

THORNDIKE PRESS
A part of Gale, a Cengage Company

Copyright © 2020 by Carl Safina
Thorndike Press, a part of Gale, a Cengage Company.

LIBRARY OF CONGRESS CIP DATA ON FILE.
CATALOGUING IN PUBLICATION FOR THIS BOOK
IS AVAILABLE FROM THE LIBRARY OF CONGRESS

ISBN-13: 978-1-4328-8307-2 (hardcover alk. paper)

Published in 2020 by arrangement with Henry Holt and Company

Printed in Mexico
Print Number: 01   Print Year: 2020

The more the habits of any particular animal are studied by a naturalist, the more he attributes to reason and the less to unlearnt instincts.

— Charles Darwin, *The Descent of Man*, 1871, p. 46

CONTENTS

CONTENTS

PROLOGUE

A flock of scarlet macaws bursts from the deep rainforest like flaming comets, several dozen big, bright birds with streaming tails and hot colors. With much self-generated fanfare, they settle into high trees above a steep riverbank. They're noisy and playful. If this is the serious business of their lives, they seem to be enjoying themselves and each other. Even within the flock, it's easy to see that many travel as close-together pairs. Following one such pair is a third bird, a hefty juvenile from last year's breeding season who is continually begging and bothering its parents. The other year-old macaws have learned a more dignified independence — if you'd call hanging upside down, fooling around, and flirting "dignified" — and have begun sorting themselves into their own young lives.

A little chimpanzee rides on his mother's back to a water hole. It's the dry season, so only scattered shallow puddles remain. It's

9

hot. Everyone has been in a distant fruit tree all morning, and after trekking through thick forest, the whole group is keen with thirst. The mother searches out some moss, wads it into a kind of sponge, dips it into a tiny puddle, puts the wet sponge into her mouth, and presses out a drink. Her little prince hops off, taps her until she gives him the sponge, and does the same. After this crucial lesson about how to quench thirst in the dry season, he and his mother relax enough to find his friends and indulge in socializing.

Meanwhile, in tropical water two miles deep, a defenseless infant sperm whale waits at the warm, sunlit surface while her mother hunts squid in night-black frigid water thousands of feet below. Like a balloon on a string, the baby follows her unseen mother. She is hearing the clicks of Mom's sonar. Nearby, the baby's aunt stands guard and waits her turn to dive and hunt. At the first signal of a threat to the baby, the whole family responds to the summons, rushing up from far beneath the indigo sea.

The stories in this book are about animal cultures. The natural does not always come naturally. Many animals must learn from their elders how to be who they were born to be. They must learn the local quirks, how to make a living, and how to communicate effectively in a particular place among their particular group. Cultural learning spreads skills (such as what is food and how to get it), creates identity and a sense of belonging within a group (and distinct from other groups), and carries on traditions that are defining aspects of existence (such as what works as effective courtship in a particular region).

If someone in your community has already figured out what's safe and what to avoid, sometimes it pays to "do the done thing." If you go it alone, you might learn — the hard way — what is poisonous or where it's dangerous. It's highly practical for members of a species to rely on social learning to get them the tried and true.

Until now, culture has remained a largely hidden, unappreciated layer of wild lives. Yet for many species, culture is both crucial and fragile. Long before a population declines to numbers low enough to *seem* threatened with extinction, their special cultural knowledge, earned and passed down over long generations, may begin disappearing.

This book is also about where culture has led Life (capital L meaning all of life on Earth, writ large) during its journey through deep time. The scarlet macaws' flaming bodies, for instance, offer a magnificent mystery: Why do we perceive as beautiful the colors and plumes that birds themselves also see as beautiful? Long before humans, Life

11

developed for itself an ability not just to perceive but to create — and to desire — what we call beauty. Why does the perception of beauty exist on Earth? This aspect of our present inquiry leads to a very surprising conclusion about beauty's part in evolution. We'll get into the amazing details as our journey unfolds. For now I'll just tell you that when I was writing one Sunday evening and I recognized beauty's overlooked role in driving the evolution of new species, the hair on my arms stood up.

We become who we are not by genes alone. Culture is also a form of inheritance. Culture stores important information not in gene pools but in minds. Pools of knowledge — skills, preferences, songs, tool use, and dialects — get relayed through generations like a torch. And culture itself changes and evolves, often bestowing adaptability more flexibly and rapidly than genetic evolution could. An individual receives genes only from their parents but can receive culture from anyone and everyone in their social group. You're not born with culture; that's the difference. And because culture improves survival, culture can lead where genes must follow and adapt.

Throughout animal life on Earth, the tapestry of genes is overlaid with more learned knowledge and information than humans have realized. Social learning goes on all around us. But it's subtle; you have to look carefully and for a long time. This book is one deep, clear look into things that are difficult to see.

We will see how, if you're the sperm whale Pinchy, or the macaw Tabasco, or the chimpanzee Musa, you too experience your wild life with the understanding that you are an individual in a particular community that does things in certain ways. We'll see that in a variable and complex world, cultures provide answers to the question of how to live where one lives.

Learning "how we live" from others is human. But learning from others is also raven. Ape and whale. Parrot. Even honeybee. Assuming that other animals don't have culture because they don't have human culture is like thinking that other creatures don't communicate because they don't have human communication. They have *their* communication. And they have their cultures. I'm not saying that life feels the same to them as it feels to you; no one's life does. I'm saying that instinct goes only so far; many animals must *learn* almost everything about how to become who they will be.

The whales, parrots, and chimps we will visit represent three major themes of culture: identity and family, the implications of beauty, and how social living creates tensions that culture must soothe. These species and many others in these pages will be our teachers. We will learn something from each that will widen our appreciation of being alive in this miracle we offhandedly refer to as the world.

By going deep into nature, looking at individual creatures in their free-living communities, we are going to get a very privileged glimpse behind the curtain of Life on Earth. Watching as knowledge, skills, and customs flow among other species provides us with a new understanding of what is constantly going on unseen by us, beyond humanity. It will help inform the answer to that most urgent of questions: Who are our traveling companions in the journey of this planet — who are we here with?

That's our present expedition. Ready?

■ ■ ■ ■

REALM ONE:
RAISING FAMILIES
SPERM WHALES

■ ■ ■ ■

They say the sea is cold, but the sea contains the hottest blood of all, and the wildest, the most urgent.
— D. H. Lawrence

Sylvia had been quiet.

Then in a private moment she turned to Shane and said, "You feel the burden of the trust these whales have placed in you."

It was something he'd always felt but could never quite put his finger on, could never articulate. In that one sentence, Sylvia explained to him why he was here.

And when he got to shore, he called his wife. She picked up and could hear in his voice that he'd been crying.

He said, "I finally understand."

And she said, "Tell me what happened."

SPERM WHALES

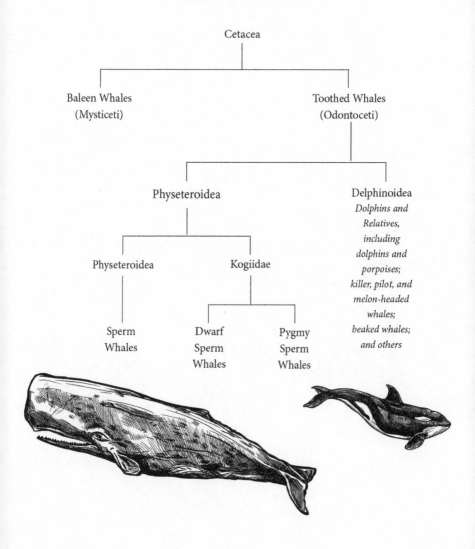

Cetacea

Baleen Whales
(Mysticeti)

Toothed Whales
(Odontoceti)

Physeteroidea

Delphinoidea
*Dolphins and
Relatives,
including
dolphins and
porpoises;
killer, pilot, and
melon-headed
whales;
beaked whales;
and others*

Physeteroidea

Kogiidae

Sperm
Whales

Dwarf
Sperm
Whales

Pygmy
Sperm
Whales

FAMILIES:
ONE

Such harmony is in immortal souls; . . . we
cannot hear it.

— William Shakespeare

At eight a.m. we are already traveling over
deep ocean. We're at what's called sea level,
as though an ocean is solely surficial, the
mere zero of altitude, as though everything
that matters rises and resides as do we, in air.
In reality we are skimming the thick, wide,
densely inhabited world beneath us. The vast
majority of the life on Earth flows through
the universe below. And that includes the
whales who share our breath but make their
lives tunneling through the sea.

How does a whale find meaning in life? This
is a very serious question that will take us far
from our comfort zone.

Already, I feel our exposure out here, at so
many mercies. Our thirty-foot boat, an open
one, is crowded with gear, crew, four young
graduate students who traffic in curiosity and
adventure, plus Shane Gero. Plus me. We're

19

running southwest into a stout chop that is building. And the captain, David Fabien, a huge dreadlocked Caribbean man with a boom-box voice, is taking these seas much too hard. I am on the boat's windward side and am soon fully drenched. I know this is his way of testing me, so I don't give him the satisfaction of turning around to glance at him. I've met far worse water and far meaner people. I figure that my taking flying sheets of seawater in stride will ensure that he and I will be good for the duration of this trip.

Meanwhile Shane is shouting, "We couldn't believe it!" Another wave showers me, and he continues: "That first month — it was the first time I really got to know sperm whales as individuals. It was just spectacular." He's telling me about his first experience here, off Dominica, in these Caribbean waters.

We soon encounter several dozen deep-flapping, dark-winged birds milling ominously. Frigatebirds. Buoyant on the wing and big, they seem forbidding and piratical. In truth, they *are* forbidding, and piratical. More formally named "magnificent frigatebirds," they are that, too.

And under the flying pirates: dark, dolphin-like fins are slicing the water. We stop. One bird hovers, then deftly plucks a squid from among the large swimming animals.

I don't recognize whose fins are driving the squid up, but Shane instantly knows them.

Genus *Pseudorca,* false killer whales — far smaller than "true" killer whales. As various individuals breathe and vanish, we guess there are about a dozen and a half of them. A long, slick patch of water informs us that we have just missed seeing some very successful hunting. Through the slick they roll their round black heads, relaxing like people after a big breakfast who don't feel like clearing the dishes.

Before we move off, Shane leans over and says, "That soaking was entirely for your benefit."

I tell him, Yes, I know.

"He'll take it a little easier from now on."

And we continue. And he does.

We seek a classic sea monster: the sperm whale, the archetypal whale of human imagination, Jonah-slurping Leviathan of the Bible, catastrophic smasher of the ship *Essex,* Ahab-maddening table-turning star quarry of *Moby-Dick.* In myth, real life, and fiction, this is the whale that looms largest in our psyches. To that almost-never-glimpsed being, so famed for rage, the world's largest creature with teeth — we now seek the closest possible approach.

For centuries, whales have *represented* things. They've represented commerce, jobs. Adventure. Money. Danger. Tradition and pride. They've represented light and food.

They are raw material, like iron ore or petroleum, from which many products can be made. And for all these things, whales have been targets. Men saw in whales everything — except whales themselves. To see things as they are requires honesty.

From this boat we seek the actual creature, living its authentic life. The mammals most specialized for water, whales descend from land mammals who slowly reentered the sea fifty million years ago. Scientists call whales "cetaceans," from the Greek for, basically, "sea monster."

Sperm whales are the only surviving members of a family, called Physeteridae, that has lasted for more than twenty million years. A dozen or so other whales of this family no longer exist. Leviathan is the last trickle of a torrent that had flowed through oceans of a richer prehuman Earth.

But at this moment we are here, and we remain contemporaries. And for the next several weeks I hope, with Shane's considerable help, to narrow the gap between us. I seek encounters that will enable me not just to see Leviathan, not just to observe sperm whales, but to penetrate past the labels and feel the beings being selves, living with their families, sharing the air where our two worlds meet. I seek merely the miraculous, and for that I am positioned in exactly the best of places: a mostly wet, hard sphere in the third planetary orbit from a star called Sun, the place where miracles come so cheap that they are routinely discarded. Hard to believe, I know. Let us proceed.

A few miles back, toward the climbing sun, steep volcanic slopes gleam emerald. The ancient Caribbean island now named Dominica helps form an arc of several volcanic isles that enclose the Caribbean Sea on their west flank and confront the open Atlantic to their east. Dominica's northern neighbor is Guadeloupe, and across its southern channel rise the peaks of Martinique. Their jungle-tangled slopes all continue plummeting right through the sea surface, meaning that deep ocean presses blue shoulders tight against these isles.

Sperm whales inhabit a wider and thicker swath of Earth than any other creature except humans, ranging the ocean from 60 degrees north to 60 degrees south latitude and from the surface to black, frigid, crushing depths. (Females and young generally remain between 40 north and 40 south.) But humans seldom glimpse them. They haunt open-ocean waters of profound deepness, almost always distant from continental shelves, seldom venturing into water shallower than about three thousand feet deep, putting them far from most coasts. Not only that, they can move forty-plus miles a day, more than twenty thousand miles annually. The scale of the theater — trackless ocean, millions of square miles — makes studying their wandering lives almost impossibly difficult. Here in Dominica, though, very deep water close to land makes this the best place in the known world for a shore-based team to attempt to reach and record them.

Shane has basically drawn a box on the ocean, twenty kilometers (about twelve miles) on a side, and said, "We're going to study one of the largest and most elusive creatures in the world when they come and go from this little box." He has invested a lot of time and effort in making this audacious proposition work. Failure isn't an option; the stakes are too high, for him and for the whales.

A curtain of light rain enshrouds us as we approach our first stop. We are hunting Leviathan, yes — but not by looking. We'd be very unlikely to succeed by just riding around and searching for a whale's blow, because sperm whales spend about fifty minutes of every hour underwater. Hunting in black and frigid depths thousands of feet beneath the waves and traveling to and from those depths occupies more than 80 percent of their time. So, like the whales, we will hunt by using water's superior ability to conduct sound. We'll listen.

We stop. A waterproofed microphone called

a hydrophone gets let down over the side. Shane's students note location coordinates, conditions of sea and sky. He passes me the headphones; we take turns listening for the clicking of sperm whales' self-generated sonar.

When meeting dolphins at sea, one may hear their squealing and whistling communication as they rapidly pace a boat or ride the bow wave. That whistling is not their sonar. Sonar comes in clicks.

Sperm whales were long thought silent. The first description of their clicks was published in 1957, by scientists. Whale *hunters* never heard the clicking sounds these whales make.

Nor do I. I hear water sloshing at the surface. It takes a few moments to get my brain to filter out the water noise, to listen deeper. Then, yes, I hear calls. Squeaks and whistles, very high. Not very loud. Shane says these are probably from the false killers we saw way back there under the frigates. Yes, the calls travel far. He says the false killers' whistles sound rather electronic; dolphins sound more breathy. Like the dolphins', their communication sounds like whistles and squeals, but their sonar sounds like streams of clicks, sometimes fast enough to buzz.

The sperm whale sonar we're searching for goes *click, click, click.* That — we are not hearing. Unlike dolphins, sperm whales also click their *communication.* The entirety of the sound they are known to generate comes in clicks, some for sonar, some for communicating.

The sea is a swirling mosaic of moving currents and seasonally shifting temperature boundaries. So inhabitants of the open ocean move continually, tracking optimal temperatures and, mainly, food. They live nomadic lives, of epic breadth and depth.

A traveler just under the surface of the open ocean may encounter little change across great distances, but a mere thirty-three feet down, the pressure has doubled. Sixty-six feet down, the pressure is three times what it is at the surface, the water is so hungry for your heat that if you were skin diving it would soon chill you, and a reduced palette of colors penetrates already-dim light.

Both ocean and land have shaped who whales are. Whales are vertebrates — more specifically, mammals. Vertebrates evolved in the ocean, mammals evolved on land, and then some returned to the sea, becoming whales. Fish bequeathed to all vertebrates our basic body plan, including our skeleton, organs, jaws, and our nervous, circulatory, digestive, and other systems. When fish brought this blueprint ashore, land and air worked to turn rudimentary limbs into walking legs and flapping wings, to turn scales into feathers and fur.

26

But when some mammals went back to the tides and immersed again, water reminded them about fins. You can sense history in whales' flippers; they merely mitten the same finger bones I'm using to type this sentence. Returning to the sea after millions of years of aboveground testing, the reimmersed mammals also hung on to: lungs, their internal heat furnaces, and parental care of their young. They packed their acute intellects and high-minded social skills into their dive bag. These attributes, developed for a life on land, confer devastating hunting advantages to the sea creatures who possess them. Seawater's oxygen content is less than 1 percent, and for animals that breathe water with gills, this has consequences for exertion. But air is about 20 percent oxygen. New adaptive retrofits notwithstanding, whales and dolphins remain every bit the mammals they ever were — and more. Quick-witted and communicative, sucking densely oxygenated air into their fast-burning musculature, they are hot-brained and über-aerated apex predators from another realm who run rings around their prey.

The sea offered returning mammals two main advantages. One: food swarms. For less-than-large creatures in the sea's open vastness, safety comes *only* in numbers. So small fishes and squids travel in crowds quite unlike anything on land. Often by the millions. Another advantage: water's sound-conducting

27

superiority. Visibility in the ocean is only a hundred yards under the best circumstances. Just a few hundred feet from the surface, no sunlight penetrates. But, being about eight hundred times denser than air, water is very friendly to sound.

When hunting, sperm whales produce sonar clicks at about two per second, a rate like: "One and two and —." "Click" is the word scientists use, but depending on distance it can sound like evenly spaced ticking; or, closer, like castanets; or, very close, like steel balls clacking.

One reason for Leviathan's absence at the moment: sperm whales don't like those false killers. One can excuse sperm whales for behaving as if the ocean is a dangerous place. They worry about real killer whales, avoid the bullying of pilot whales, and act touchy about false killers, who harass young sperm whales by biting their tail flukes, seemingly for fun. It's not fun for the sperm whales, who are shy and who take good care of their children.

Shane consults the GPS for his next position. We go the three kilometers to get there. Our listening devices can detect a sperm whale's sonar over at least five kilometers. So our listening stops are spaced to leave no sound gaps. If whales are present, we will detect them. If they are not present, the silence will speak volumes.

We know enough about whales to fill many books. But we know a very small fraction of what their lives entail, of how they experience it. Sperm and humpback whales, killer whales, and bottlenose, spotted, and a very few other dolphins have been the subjects of sustained human inquiry. Most whale and dolphin species, dwelling in their liquid world beneath the planet's curving blue horizons, remain near-total strangers to us. Every few years, scientists discover the existence of a previously unknown whale species.

Proximity to Leviathan is more easily imagined than accomplished. The farther offshore, the choppier, wetter, and less comfortable. The sea is under no obligation to yield its whales easily.

But Shane Gero is driven. Trim and athletically built — a lifeguard's physique — with short brown hair and blue-gray eyes, he combines a likable, undefended friendliness with a deep-ranging, inquiring mind. Shane has made these questions his quest: How does a sperm whale learn who he or she is? How do sperm whales teach their children to wield the codes of their identity? Getting answers would reveal how sperm whales construct their remarkable sense of family.

The second listening stop was quiet. As we head toward the third, the sea reflects a glar-

29

ing haze that scatters light everywhere. The distant island, Dominica, fades in and out of blanketing clouds. Moving across the sea surface, we feel as if we're skating over mysteries beyond human ken. We are.

Our zooming hull scares up flying fish; one lands in the boat. I admire its large eye, mirror sides, and the indigo streak along its back. Then I flip the fish into the sea.

Just past that patch of flying fish, a white-bodied, streamer-bearing red-tailed tropic-bird comes from who-can-say-where and begins tracking us. The bird *knows* that our sea-slicing hull can scare flying fish into the air, *understands* what might happen, waits with *expectation* for what it has *in mind*.

We disappoint. I look up, and the bird looks at us, and I think, "Why weren't you here five minutes ago? We were scaring up plenty."

Approaching the third eavesdrop, we encounter a quarter acre of floating yellow-green sargassum seaweed. From it we pluck a plastic tarp. A small group of dorado — first revealed only by electric-blue pectoral fins moving through the deep dark of the sea — come to our hull. With neon fins and oar-shaped, arm-length bodies speckle-decorated in blues and yellows as if colored by a child, they are perhaps the most beautiful fish in the full phantasmic panoply of all fishes.

■ ■ ■ ■

On our third stop, the hydrophone again descends on its wire into the planet's fluid wrapping. I hear an engine. But wait: "That ship engine is so loud, I —"

Shane thinks he hears very faint clicks. "I'm not sure —"

Now we can make out barely discernible electric-sounding whistles. Shane's unsure who. I'm bewildered by the complex subtleties.

Then, something else. Through the sound of surface noise and the sound of the distant ship, through the whistles —. Clicks.

Sperm whales click. But certain dolphins, too, can click. And now we see traveling dolphins coming upwind in the glare, into the wave tops, glistening out of the ocean in little groups.

Who are we hearing?

Shane listens intently, headphones on, eyes closed, trying to filter a click out of the sounds of the ocean. To cut some of the noise, he immerses a "directional" microphone. It's just a hydrophone inside a salad bowl on a broomstick, a comedy of high tech meeting highly improvised. The bowl shades the mike from sounds other than those coming from the direction it's facing. Turning the stick localizes sounds. It's as close as we can

31

come to cupping our ears underwater.

"Not nearby. That's for sure."

I gaze at the slaty sea. It's hazy, glary, breezy, choppy. Bleak.

Turning the stick on the directional mike, his hat brim low, his attention focused on listening, Shane quietly says, "Yes. Could be four, maybe five whales —." He pauses while continuing the turning of the salad bowl. "One's northeast. Most are south of us."

In high anticipation, we look south. It's very rough there. We go a short, wet distance. Shane isn't eager to pound and slam into the windy, soaking chop.

Our day is about whales. About finding whales. About identifying who we find. And these vague clicks are how this day begins allowing us to peel its secrets. Far below and far away, sperm whales are hunting, clicking to determine what's ahead in their darkness.

Leviathan inhabits — and creates — a world of sound. Almost constantly, whales hear the sounds of dolphins, of other whales, and of their own family. Almost constantly while they are deep underwater, they generate and listen to sonar clicks.

Jacques Cousteau famously titled his 1953 book *The Silent World.* It's an evocative turn of phrase — but far off the mark. The sea shimmers with callings and affirmations. Warnings. Hellos. Yearnings of love-desire.

Tribal chants. Engines, pneumatic air guns, and the thrumming of what's coming. Since water is eight hundred times denser than air, sound travels four times faster through it, making water a superb medium of communication. That's why so many animals, from shrimp to whales, have developed ways to make the sea transmit their aural messages. Some — pistol shrimp, mantis shrimp, possibly certain dolphins — may use sound as stun guns. Because water densities vary widely with vertical bands of layered temperatures and salinities, oceans become acoustical transmission systems allowing properly tuned sound to reflect across layers of seawater and travel longer distances, a bit as radio transmissions can travel farther by going from one repeater tower to another. That's how blue and fin whales, booming at the lowest frequencies, can stay in contact and travel "together" while spaced across hundreds of miles. Anything but silent, the ocean is brimful of sound and messaging.

The sonar of sperm whales is the most *powerful* burst of focused sound made by a living thing. At around 200 decibels, sperm whale sonar is one of the loudest sounds known. The whales concentrate a cone of energy ahead of themselves. That our equipment can detect it across three miles in any direction from the whale means that the whale is literally causing the vibration of several cubic miles of seawater, a vast high-powered sphere of sound, an extraordinary envelope of energy.

So powerful and penetrating are their sonar clicks that sperm whales can likely see what many things look like inside, as if X-raying them. Humans who slip into the water near sperm whales sometimes get scanned with rapid bursts of audible clicks that can be felt as vibrations. Of a young orphan with pneumonia who had beached himself while weak and near death, Richard Ellis wrote, "He made a 'pop' so loud that it knocked my hand off his nose."

When we stop, Shane again submerges the directional hydrophone and immediately announces, "Some to the north."

Our pace accelerates. It feels like a hunt.

After several kilometers of determined northward travel, we stop. And this time, clear and even, I distinctly hear a sound like a fingernail slow-tapping on a hard countertop.

Sperm whales. Definitely. But briefly. The taps stop. Why — ?

"They might be heading up."

When sperm whales stop hunting, they stop making sonar clicks and simply begin a long ascent toward the sun to replenish their breath.

Shane insists that, considering the time that

34

has passed since the clicks stopped, we should be seeing the blows of at least one whale at the surface. But the mirror-ball glare across the rough and white-streaked sea could hide a whale.

We stare into the dazzling chop, searching shimmering shards of brightness for evidence of breath. The boat rocks. The sea rolls. The ocean is pure glitter.

The headphones report faint and distant clicks far to the northeast.

Shane says, "Wow, they're just super spread out today."

But the whales can easily hear each other. For them, being able to hear their family members counts as "together."

"Okay," Shane instructs. "Let's go northeast, try to pick up the main group, see who these are."

Shane was the kind of kid who raised tadpoles in kiddie pools and watched caterpillars become butterflies. By eight, he wanted to be a marine biologist. When he was twenty, he saw a wild whale. Awestruck by that experience, he wrote an e-mail to *the* pioneering sperm whale researcher, Hal Whitehead. Many weeks of waiting followed. No answer came. Then Whitehead responded, and Shane's life changed.

Before Shane and Whitehead first sailed into these waters, Dominica was rumored to have "resident" sperm whales. Whitehead had documented a nomadic existence for Pacific sperm whales. Scientists had never observed "resident" sperm whales. He and Shane were skeptical.

But in the first hour in these waters, they met a whale family they named Unit T. Later they met whales they dubbed the Group of Seven — and spent an unprecedented forty-one consecutive days with them. Soon they met half a dozen other whale families. In the brief history of researchers studying these mythic giants, no human had ever been granted such intimacy.

Now we hear by their sudden silence that the whales who'd been to our northeast are coming up. With the pressure, temperature, and light changing drastically, the dissolved gases reexpanding into their collapsed lungs, from a world we do not know they rise up toward the planetary curve, the surface of the sea. The ease of the surface, the warmth we know, the air we share.

Captain Dave announces, "Blow!" And Shane shouts, "Yesss!"

About two hundred yards away, a whooshing left angled puff of steamy gray breath jets from a massive machine of a head, a gigantic ocean-splitting wedge that spans a full third of her entire length. Unlike all other whale species, a sperm whale's blowhole is not atop the head but at the tip, where you'd expect nostrils on a typical mammal's snout. A muscular mound controls the opening and clamping shut of the weirdly left angled single nostril.

The wet vapor dissipates in the breeze. She cycles several more breaths. Breathe. Ten or twelve seconds pass. Breathe. Another twelve seconds or so. Breathe. Breathe for the minutes it takes to purge and to recharge so many barrels of oxygen-freighted blood. Because their lungs collapse from the pressure of deep dives, sperm whales are powered not principally by lungs full of air but by oxygen prepacked into muscle.

We pound closer for a better look. From fifty yards distant, she moves toward us. The skin of her head is taut, like dark shrink-wrap. The rest of her body is wrinkly, to reduce water drag by breaking the laminar flow. Her eyes, of limited use in black and frigid depths, are relatively small. Her size makes the hourly depths-to-surface round-trip a realistic proposition. Her sonar overrides the blackness. Her blubber defeats the cold. All her extremes are perfections.

She blows and then, dipping that massive muzzle and bending her long back, she flags her imminent departure from sunlight and air by hoisting aloft her wide black propeller. Shedding sheets of water, her flukes and sturdy tail stock shove her downward into the swallowing sea, toward hunting depths perhaps a hundred body lengths down.

"Well," Shane says inconclusively. "Interesting."

And what I am left with is this impression: a whale is too big to see. At a time, you get pieces. Now the head. Now the back. Now the flukes. Never the whale. In Rome once, I said to my wife, Patricia, "We've now seen Michelangelo's painting of the Creator. But what would the Creator's own painting of creation look like?" I think that is easy to answer now: it is these whales, in this sea.

"She just focused her sonar on us," says Shane, still listening. "Now she's traveling down." Focused sonar comes in very fast strings of clicks called "click trains," sometimes more than six hundred clicks per second, and they sound to us like a buzz.

Captain Dave says, "She look like a teenage whale, y'know."

"Yeah, not a big one. But I don't think that's who we were first hearing."

This informed guesswork usually spirals in on the correct identity.

This moment the question remains: Who? Which family?

Suddenly, four hundred yards from us a different whale blows. She's moving steadily, a dark shape beating a white path through the waves.

We amble toward this new whale. Every ten

38

seconds or so, she blows short puffs of steamy breath, purging and packing in a fresh fill.

Suddenly and just a boat length away, a whale only about fifteen feet long pops up.

Shane yells, "*Neutral! Neutral!* Mom's right here!"

I look down and am astonished to see the dark visage of a huge whale. I have a hard time understanding quite what I'm looking at.

"She's vert-sleeping," explains Shane.

Now I realize: the mother is resting vertically in the water, nose up. I can't quite see her tail; she's too long. It was *her* blow that first got our attention.

When sperm whales sleep, they sleep vertically. "They kind of bob up to breathe," Shane says. Whales always have to breathe intentionally; it's never automatic.

The small whale is making short, shallow dives. Shane says, "The youngster will go down and rub the mammary area to induce the letdown of milk. That's usually how we tell they're nursing."

Milk is the mother in liquid form; the growing infant mammal is entirely its mother's milk transformed into flesh and blood, bone and all the growing organs and systems, the pulsing and purring. Most sperm whales nurse for four or five years before weaning. Some, much longer. In these waters, the oldest juvenile who ever appeared to be nursing was eight years old. The known record is thirteen years. Mothers don't get pregnant again until they've stopped nursing. They can live to about age sixty-five, but the oldest known pregnant female was around forty-one years old.

Nursing habits differ among different families. In the Group of Seven, young ones suckle only from Mom. The J's are a family prone to communal baby nursing. In Unit T, Tereka, who has never been known to give birth, helped nurse two young ones named Top and Turner. "Sometimes both nursed at once, on each side of her," Shane recalls. One at each teat. "Which is amazing."

How could that happen? Why?

"That's just how the T Family does it."

In the Group of Seven, Digit was three years old, and had stopped nursing when she got tangled up in some fishing gear, which slowed her down. After Digit's movement was compromised, her mother, Fingers, began nursing her again. Now at age six, Digit is still dragging the gear around — and is still being nursed.

For sperm whales, family is everything. In Shane's early years here, the family he'd dubbed the Group of Seven liked to hang with a family called the Utensils. An adolescent female of the Utensils named Can-opener liked to play with the Group of

40

Seven's young ones of that time, named Tweak and Enigma. Since Digit's tangling, the two families have continuously remained together as if they are one. Is it because they recognize that Digit is in distress? (The rope tangling the base of Digit's tail will likely cut into her as she grows, and probably kill her. But Digit is still moving too fast for a human rescue attempt.)

Families such as these who particularly like one another are called "bond groups." It's a designation borrowed from elephant researchers, and it refers to families who are good friends. In fact, the social structure of sperm whales more closely resembles that of elephants than of other whales. The parallels are many: tight, stable families of females and dependent young; bachelor groups of physically mature males who defer breeding for years rather than compete with colossal older males; striking size disparity between females and full-blown males; the largest brains in their medium; even their ivory-bearing teeth. Female elephants and sperm whales are sexually mature by their early teens. Females of both remain in their birth family for life and give birth into the same family in which they were born. Male elephants leave their mother around adolescence. Sperm whales, same. Sometimes sperm whales from a family will, for a few hours or days, travel with other whales and then split up. Elephants do something similar. I've watched herds of elephants that seemed like one big group, sometimes numbering in the hundreds. But then at day's end, much smaller groups would split off. These were families headed into the hills to the places where they'd spend the night. When families merged, I could not keep track of who was who. But the elephants knew who they were with as surely as you know your own family when you're with them in a crowd.

The whales at the surface now could stay up for just a few minutes or linger longer. Though a typical hour of a sperm whale's adult life consists of a long dive — traveling down to food, searching for food — between surface breathing intervals of ten minutes or so, sometimes something else happens. "Once in a while they just decide to chill," says Shane, "resting and socializing for a few hours."

The young one goes into a dive. Young whales don't usually do that, but —

"Codas!" shouts Shane.

Short bursts of clicks are coming through the headphones. "Codas" are not the steady ticking of whale sonar. Codas come as varied rhythmic patterns, a little like simple Morse code, of three to as many as forty clicks. They are sperm whales' signals of identity. Declarations of belonging. By signaling with their codas, they announce themselves, determine the identity of other whales and whether they've encountered a group they can socialize with — or must avoid.

The whales often make codas during transitions, such as when going to the surface or about to dive, when greeting family members, when a male is present, or if they've detected predators. When the pioneering researcher Hal Whitehead first witnessed a whale delivering a new baby, he wrote, "There was a particularly heavy burst of codas at the time of the birth."

Through the headphones comes an aural tapestry of deep and distant sonar ticking and these nearby loud clacking codas. *So* loud, it's like someone clapping right next to my ear. In fact, as soon as I put the headphones on, I thought it was Captain Dave smacking his hands together behind my head, messing with me. I'm amazed at the clarity and strength of their talking. They're going: "One. Two. Three-four-five."

No one fully understands what information is coded into those patterns. Except, of course — all the whales do.

The mom who'd been sleeping is now exchanging codas with another whale. There is a back-and-forth, a call-and-response: "I am here," one says. "And I am over here," says the other. A conversation of sorts. Likely the youngster had dived because she went to greet whomever is approaching.

And now, yes, another adult female appears with Mom. The three whales rest at the surface side by side, breathing. Youngest on the right, the largest on the left. I sense deep relaxation after deep exertion. Each forceful, intentional exhalation wafts faint rainbows into the breeze.

The two larger whales slide below for a bit of vertical rest. After just a few minutes, the small whale begins tail-slamming the surface, detonating impressive splashes. Even a small whale has a big, hard-slapping tail. The youngster seems to want action. We count twenty-one slaps.

"He's like, 'C'mon, Mom, wake up,' " Shane laughs.

Meanwhile, Shane has been listening to two other whales, who sounded like they were a couple of kilometers away. They have fallen silent.

Moments later, yet another whale explodes full-body out of the ocean about three hundred yards away, launching with attitude, back slightly arched. And as she crashes back, she seems intentionally to slam her head onto the sea surface for maximum effect. On a second vault she bursts into the air with jaw open wide, water streaming from the corners of her mouth. The once-in-a-lifetime image immediately photographs itself into my memory. She goes down some distance, turns

44

toward the light, then pumps her broad tail with force sufficient to propel her vast body into the air. Through four more breaches, her strength and mass astound us.

Her name is Jocasta. The edges of the whales' tails are remarkably irregular. They bear scratches, gouges, and bites, perhaps from sharks or false killer whales. Such lesions and lacerations heal, but the flesh doesn't grow back into its former shape. The set of woundings life inflicts on each whale are unique enough to allow positive IDs. Jocasta has two scallop-shaped marks. These whales are hers, the J Family.

FAMILIES:
TWO

At some point long ago, a group of sperm whales found plentiful food up against the deep-plunging shoulders of these islands. If they could recognize that "it would pay for us to stay around here," as Shane says, and "to, in a sense, agree on that," then communicating such a thing among themselves would aid their survival. Before they could come to such an agreement, however, they needed a way to understand who "us" referred to. They needed a way to understand who they were, whom they belonged with — and which whales "us" left out.

Somehow, sperm whales — worldwide — developed the capacity to understand group identities, distinguishing families and groups of families called "clans." The way sperm whales recognize and announce identities and group membership is with their codas. Sperm whale babies go through a babbling period similar to human infants' (as do the young of monkeys, apes, dolphins, and some birds).

By the time they're a couple of years old, though, they've acquired the family's codas. As young children learn the language of the group they are born into, young whales learn the codas of their family and clan. The whales are always individuals within families, carrying on their lives in detail that is as vivid and present to them as ours is to us. Across the open ocean and in abyssal darkness, over long decades, "what they have," Shane has come to appreciate, "is each other."

Shane's long glimpse of the whales of these waters has led him to a thought both obvious and profound: they have lives. For centuries, the only thing humans cared to understand about whales was how to kill them. Almost too late we acquired a modicum of respect. So recent is this new curiosity that Shane Gero is among the first to ask of whales, in essence, "What is life *like* for you?"

"When we're home paying bills and taking care of the kids and at work," he notes, "the whales Raucous, Roger, Riot, and Rita are hunting for food or caring for the baby, communicating and resting, and, mainly, staying together. Life for them is always going on, parallel to ours."

About two dozen families of sperm whales pass through these waters. Some, Shane has seen only once in his fifteen seasons in these waters. About sixteen of those families come and go regularly. Of these, Shane knows ten families intimately enough to recognize them on sight by the shapes of their tails.

So far, Shane has spent about six hundred days in direct contact with sperm whale families here. That's to say nothing of time spent in prep for studying sperm whales, the logistics of studying sperm whales, mentoring and teaching about sperm whales, writing proposals to fund sperm whale studies, computer analysis of sperm whale sound, graphic production of sperm whale tracking data, writing, revising, and working through the peer-review process, and publishing his work about sperm whales in scientific journals. If being a sperm whale means being immersed in all things sperm whale and being focused on your family, Shane Gero comes about as close to being a sperm whale as a human can get. What he's achieved is that the culture, social lives, communication, genetics, movements, and diets of these sperm whales are better known than those of any others in the world.

"I'm under no delusion that the whales know who I am," Shane explains. "But I've put a lot of time and effort into learning who *they* are." And basically, "despite the great differences in their physical surroundings compared to ours, everything we've learned about them so far is very" — he searches for the right word — *"relatable."*

48

The story here, Shane says a bit mysteriously, "is about different definitions of a concept of 'us.' " More simply, he adds, "The main thing I've learned from whales is that your experience of the world depends on who you experience the world with. Who you're with makes you who you are. The biggest take-home about life as a sperm whale is: your family is the most important thing. If I ever prioritized working with the whales over being with my family," he continues, "I would have failed to learn the main lesson the whales have taught me: learn from Grandma; love your mom; spend time with your siblings; share the burden of what's needed. Spending so much time with the whales has changed how I value the people in my life. The life of a sperm whale is complicated and joyful and hard, and the kinds of things they experience are comparable — in *their* way — to things we go through and worry about. Trying to learn what the whales value has helped me learn what I value. Trying to learn what it's like to be a sperm whale, I've learned what it's like to be me."

I tell Shane that he sounds like Captain Ahab after twenty years of psychotherapy. One senses that as much as the whales interest him, they also haunt him. He is committed to pouring all his thought and effort into pursuing these whales for the rest of his life. He seems deeply affected by humanity's previous, and ongoing, relationship with whales. And though he is braced for what might be coming, he has no way of knowing how any of it will go.

Moments unwind. Shane seems finished with his train of thought. And then he adds, slowly and with great emphasis: "We need" — pause — "to find ways" — pause — "to coexist. These are rich, complicated lives being lost every year. No one notices. And that is painful, because, *I know them.* I can't expect everyone to know them. I can't expect anyone else to spend thousands of hours in the company of sperm whales. And in exchange, my obligation to them is to make it mean something. I struggle with how to make people care. I've had fifteen years so far to try to figure out what it means to be a whale. They've helped me understand how I can be a better human being. Now I have to figure out, What am I gonna do — *for them?* That keeps me awake."

When we think of culture, we first think of human cultures, of *our* culture. We think of computers, airplanes, fashions, teams, and pop stars. For most of human cultural history, none of those things existed. For hundreds of thousands of years, no human culture had a tool with moving parts. Well into the twentieth century, various human foraging cultures from the tropics to the

Arctic retained tools of stone, wood, and bone. We might pity human hunter-gatherers for their stuck simplicity, but we would err. They held extensive knowledge, knew deep secrets of their lands and creatures. And they experienced rich and rewarding lives; we know so because when their ways were threatened, they fought to hold on to them, to the death. Sadly, this remains true as the final tribal peoples get overwhelmed by miners, loggers, ranchers, and planters who value money above humanity, which is perhaps the most salient characteristic of our culture. We are living in their end times and, to varying extents, we're all contributing to those endings. Ultimately our values may even prove self-defeating.

The value of cultural diversity in the human family has been under-appreciated. Many cultures have been lost. The importance of culture in the other-than-human world has been almost entirely missed. Human recognition of this lapse is in its infancy. For the last thirty years or so, the diversity of all living things, "biodiversity," has been thought of as operating at three main levels: the genetic diversity within each species, the diversity among species, and the diversity of habitats (grasslands, forests, deserts, oceans, and so on). But there is a fourth level in living diversity, and it is just now becoming recognized: cultural diversity. Culture comprises knowledge and skills that travel from individual to individual and generation to generation. It is learned *socially*. Individuals pick it up from other individuals. It is knowledge that doesn't come from instinct alone. It's not inherited in genes. What is learned and shared: that is culture. Our understanding of living diversity is only just beginning to recognize that what is learned and shared is often crucial to survival.

A baby sperm whale has a lot to learn. Deep-diving skill develops over time. Youngsters learn by accompanying their mother and other adults. Before they can use their own sonar, they likely "eavesdrop" on information in the returning echoes of the sonar of adult family members, assimilating the sounds of detection, and of pursuit. The answers to other questions must also be learned. Where in these currents and underwater slopes is the hunting best? How do we travel; where do we go in changing seasons? Perhaps *all* of that must be learned. Perhaps, also like elephants, they rely on the oldest individual's knowledge of where to head during a time of food shortage. But how can we humans recognize what whale behaviors and skills exist only because they are learned *from* others?

What is cultural becomes obvious if not everyone does it. Everyone eats; eating isn't culture. Not everyone eats with chopsticks; chopsticks are part of culture. All chimpanzees climb trees; that's not cultural. Some chimpanzee populations crack nuts with hammer stones and anvils, but not every population that lives where there are nuts cracks them. It's cultural. Group-to-group variations in customs, traditions, practices, and tools — such differences show what is cultural.

When a devastating drought hit East Africa in 2009, hundreds of elephants died. Survival was much higher in families who had a matriarch old enough to remember how her family had survived the previous severe drought, more than two decades earlier, and led her family to remaining water. Of fifty-eight elephant families in Kenya's Amboseli region, one family lost twenty members, while the "KA" family did not lose anybody. The KAs were led by two big females — Kerry and Keira — who were forty and thirty-nine years old when the drought hit. "Old enough for wisdom," as the legendary researcher Cynthia Moss described them to me. The KAs spend a lot of time north of the park. "It is not particularly safe," Moss says of the area. "However, they are obviously doing something right, and I would attribute that to the knowledge of those two old females. In fact, between 2005 and 2019, a span of fourteen years, they lost only one baby. Truly amazing." Like sperm whales, elephants live in families where age and

53

experience matter. They learn from their elders where to go when a crisis strikes. Without receiving knowledge from older keepers of knowledge, they'd die. That is culture.

Before the 1960s, many people thought dolphins were little different from fish. Starting in the 1960s, Ken Norris showed scientifically that a key to being dolphin is flexible learning. Flexible learning leads to different behaviors from group to group. By the end of the 1980s, Norris and others were viewing the group differences in dolphin behavior as "clearly cultural."

Local habits or particular traditions can keep individuals together — and can also keep groups apart. Community markers that enforce difference can even keep groups on hostile terms. Humans excel at this (think of languages, flags, uniforms, and so on).

For a long time, cultural separation was believed to be "uniquely human." But we're now learning that humans are not the only creatures who use signals to determine group identities, reaffirm membership, reinforce differences, and cause distancing. Sperm whales, pilot whales, orcas (killer whales), and various dolphins can tell by sound which pods they might warmly greet — and which they must avoid. Elephants know which families they like and which they prefer to avoid. Elephants, primates, and other species know who's in their group and who is an outsider. Thousands of species of birds recognize their mates and neighboring territory holders, and vigorously repel other intruders. Apes' responses when meeting other groups range from murderously violent (chimpanzees) to frisky and frolicking (bonobos). Wolves amid the chaos of pack-against-pack violence have no doubt about who's on their team and who's not. They don't wear caps with team insignias, but they know who's on their side (their family members) and who are their rivals. Group identity and self-identification with a group have long been considered defining hallmarks of human culture. But in reality they're not exclusively human.

The Pacific Northwest's so-called northern- and southern-resident orca communities' only humanly discernible differences are their vocal dialects. Both communities of orcas specialize in hunting salmon, and no apparent physical or genetic differences characterize membership between those two communities. They seem to share everything — including a disdain for the community that is not theirs. Orca communities avoid mixing for purely *cultural* reasons. Their self-segregation of stable cultural groups was until recently considered so exceptional that researchers said it has "no parallel outside humans." But it's turning out that cultural identity and cultural segregation might be more widespread than previously suspected.

In addition to those we've mentioned, certain bats, birds, and many other creatures recognize individuals by the sound of their clicks, howls, trumpets, and songs — their voices. Because voice *represents* the individual, voices are *symbols* of identity. Alarm calls are likewise *symbols* that alert to the appearance of foes or identify a dangerous predator. For a symbol to work, one must have a *concept* of what the symbol represents. Apes, monkeys, and birds have alarms for things like snake, hawk, and cat. The alarms are essentially words for these different dangers, telling companions whether to look up, look down, or climb a tree. Other species are not supposed to use or create symbols. Tell that to *them*.

For sperm whales, it takes a village to raise a child. The need for reliable babysitters seems the main reason sperm whales live in stable groups. No other large whales — humpbacks, blues, fins, grays — live in groups where the same individuals stay together for decades. Sperm whales generally live in the orbit of mothers, in a community of related whales, structured for the care of young ones. It's been called a "mother culture" by Shane's professional elders, the pioneering whale biologists Hal Whitehead and Luke Rendell.

It could as well be called a babysitting culture.

Infant sperm whales must swim; they might travel forty or fifty miles in their first day of life. However — crucially — baby sperm whales seldom deep-dive, and *do not accompany* their mothers down into the great, black, frigid, high-pressure depths where adults spend most of their time hunting. Babies often trundle along at the surface in the direction of sonar clicks coming from foraging adults far below or turn in circles while awaiting the ascent of their elders. Alone at the surface, they are quite defenseless. Killer whales come along only rarely, but when they do, they pose mortal danger for baby whales.

The solution has been for baby sperm whales to live with aunts and their grandmother in addition to their mother, all in constant contact using sound. In families with a very young baby, adult diving often gets staggered, so that when some are down, at least one adult is up nearby, monitoring the vulnerable youngster. Sometimes a young whale at the surface *seems* alone. But any distress signal brings a relative right up. If the danger is real, the whole family is quickly summoned.

"When killer whales show up, or something goes wrong at the surface, it seems like sperm whales suddenly come popping up out of nowhere," Shane says.

Other large whales solve the problem of newborns quite differently. Most large whales give birth in shallow tropical locations. In those relatively safe areas, mother whales bring their new young into the world and continually stand guard. The catch: in those warm, safer waters, there is no food for the mothers, who do not eat for several months. Most huge whales live by eating tiny things, ranging from rice-sized copepods to little dense-schooling fishes that exist in immense swarms. The dinosaur-dwarfing blue whale gulps shrimplike krill the size of your little finger, straining them from the ocean by the millions. Krill and the vast swarms of small fishes that fodder those whales live in cold climes. So each year, mothers migrate to strike a trade-off between eating and a safe nursery. Gray whales who spend the foraging season in the Bering Sea migrate to give birth in the safety of warm Mexican lagoons. Right whales travel from the Gulf of Maine to Florida. Blue whales off Alaska swim to waters off Central America and give birth there. This is life for many whales. For New England humpbacks, winter means the Caribbean. Some humpbacks migrate from the Antarctic Peninsula to Costa Rica, traveling more than five thousand miles (8,300 kilometers) and crossing the equator before coasting to a stop.

Sperm whales do it differently. They give birth where their food is. Sperm whales' main prey animals, squid, live amply in warmer latitudes. So female sperm whales have no need to fast or to migrate. The hitch for them? The squid are two thousand feet down, and infant sperm whales cannot follow. Consequently, a mother sperm whale spends five-sixths of her time far from her baby. And this dilemma, more than anything, drives the sperm whale's elephant-like social arrangement of living in female-led families where everyone knows each other and everyone protects the young.

Sperm whales who are consistent companions often stay together for life. Groups of constant companions are called a "social unit." These are sometimes close relatives, sometimes not. Their bonds are long-lasting across time and distance. During former whale-hunting times, several groups of females who'd been tagged together remained together up to ten years later and hundreds of miles distant. We know this because they were killed, together.

Their strong family bonding was first documented by Thomas Beale in his groundbreaking 1839 book, *The Natural History of the Sperm Whale.* Beale, a keen observer who got to know these whales from the deck of a whale-hunting ship, wrote, "The females are very remarkable for attachment to their young, which they may be frequently seen urging and assisting to escape from danger with the most unceasing care and fondness." He added:

> They are also not less remarkable for their strong feeling of sociality or attachment to one another; and this is carried to so great an extent, as that one female of a herd being attacked and wounded, her faithful companions will remain around her to the last moment, or until they are wounded themselves. . . . The attachment appears to be reciprocal on the part of the young whales, which have been seen about the ship for hours after their parents have been killed.

Because foraging specialization drives sperm whale family structure and, thus, sperm whale culture, we might pause to consider how odd their eating adaptations are. We've already mentioned the extreme depth, darkness, and cold in which they forage — mainly for squid — using their superlative sonar. Let's consider how they actually eat. We must speculate, because no human has ever seen a sperm whale catch a squid.

The largest creature with teeth, Leviathan can clamp onto a squid the length of a living room, squid species bearing names such as "giant" and "colossal." Battles ensue. But *most* squid they eat, such as the diamondback squid here off Dominica, are around three feet long. And many are much smaller. A male sperm whale killed off Madeira in 1959 had four thousand squid mandibles in his stomach. Of these, 95 percent were from squid that had weighed under about two pounds. It is almost inconceivable that a sixty-foot whale could get enough calories by chasing such relatively tiny squid one at a time. And many squid found in the stomachs of sperm whales didn't have a tooth mark on them.

To imagine how sperm whales eat, consider their strangely long and exceedingly narrow jaw. The jaw of all other large whales, as in most mammals, including humans, is about as wide as the rest of the head. Exceptions in mammals — anteaters, for instance — indicate extreme specialization. Sperm whales possess an exceptionally thin and narrow mandible. For most of its length, the two sides are fused into a rod. That rod-like jaw is topped with curved, finger-length teeth. The teeth are stout and round, like fat ivory carrots, with no cutting edge. When the jaw is closed, the teeth snug into sockets in the mouth's toothless upper rim. The narrowness of the mandible strongly suggests a special hunting technique. The closest thing I know of is the bladelike mandible of birds called skimmers, who snap up fish by the astonishing technique of flying along with their open lower bill slitting the surface of calm water.

All this led me to wonder whether the sperm whale jaw can double as a squid rake. Native Americans of the Pacific Northwest sometimes raked densely schooling herring into their canoes, and people in the northeastern United States used eel rakes. But for my hunch to be viable, I'd need to know that sperm whales can drop their jaw open to something approaching ninety degrees. When I ask Shane whether they can open their jaw pretty wide he says, "Yes, very. Almost perpendicular to the head." Researchers have determined that a sperm whale's clicks could detect a foot-long squid from up to a thousand feet away. Such small squid often form dense swarms. I imagine the whales rushing their great bulk through a school of squid, jaws open wide to snag or injure prey for an easy gulp.

Turns out, in the early 1800s Thomas Beale imagined something similar. Though there seems no way he might have seen a sperm whale feeding, he wrote confidently,

When this whale is inclined to feed, he descends a certain depth below the surface of the ocean, and there remains in as quiet a state as possible, opening his narrow elongated mouth until the lower jaw hangs down perpendicularly, or at right angles with the body. . . . The teeth of the sperm whale are merely organs of prehension [grabbing], they can be of no use for mastication, and consequently the fish, etc. which he occasionally vomits, present no marks of having undergone that process.

However sperm whales do it, they catch enough squid to keep them going.

What they *don't* eat — is people. Previous authors wrote nonsensically of sperm whales as monstrous gluttons seeking human flesh. Beale described them — correctly — as "remarkably timid, and readily alarmed." He added,

We might be led to believe that there is no animal in the creation more monstrously ferocious than the sperm whale. . . . [Yet] if these huge but timid animals happen to see or hear the approach of a ship or boat, their fear in all cases is excessive. . . . The sperm whale in reality happens to be a most timid and inoffensive animal . . . readily endeavouring to escape from the slightest thing which bears an unusual appearance . . . [and] quite incapable of being guilty of the acts of which he is so strongly accused.

Perhaps most astonishing for his time, Beale

put the sperm whale on trial for murder, administered a fair proceeding, and acquitted the accused on the basis of self-defense:

> The blow of the harpoon . . . when inflicted, often paralyses the largest and strongest of them with affright, in which state they will often remain for a short period on the surface of the sea, lying as it were in a fainting condition; . . . they rarely turn upon their cruel adversaries, for although men and boats are frequently destroyed in these rencontres, they are more the effect of accident during violent contortions and struggles to escape, than from any wilful attack.

Leviathan the spermaceti whale combines in epic proportion the size, prowess, and myth of all whales everywhere. It is not the very largest, yet while it does not exceed it excels, having made on human minds an outsized impression as the only great whale with teeth, the largest of the deep divers, a whale as family-oriented and inoffensive as elephants yet capable of elephantine rage, a ship-sinking defender of its females and their children. Among all whales who've written their stories throughout the seas, theirs is the magnum opus.

And so here Shane and I are in these waters, seeking a murderous monster of legend who is in reality timid and bonded to family, who treats its children "with the most unceasing care and fondness." These thoughts — of these beings, with their enormous bodies and minds, keeping their families close as they conduct their lives somewhere in another universe on our troubled planet — are quite enough to disquiet me.

Sperm whales have a distinction, and it's major. As Mark Moffett points out in *The Human Swarm,* in no species can an individual recognize total strangers as members of their society while other strangers are pegged as outsiders — with two exceptions: humans and sperm whales. A *society* is a group in which all individuals recognize other individuals as being either members of their group or outsiders. In almost all species, this limitation keeps societies small, because it requires all individuals to know every other individual in their society. Individuals they recognize are in; strangers are out. (Some social insects, such as ants, also qualify, though rather than recognize friends or foes through a combination of cognition and judgment, the insects respond more simply and automatically to chemical cues.) The ability to perceive that certain total strangers are in your clan — and thus okay to socialize with — is a truly exceptional capacity of sperm whale culture.

Across vast Pacific meridians, sperm whale

families (females and young) have formed themselves into five clans composed of thousands of whales. Their sense of identity and membership within a particular clan is reflected by their dialect of clicked codas. In the Pacific, any *single clan* of sperm whales, built of many hundreds of family units spanning many horizons, can be populated by as many as ten thousand individual whales. At those population sizes, most whales within a clan are not closely related and don't know each other. Yet all members of a clan may socialize. Members of one clan never socialize with members of a different clan. Because the different clans' nomadic movements can overlap, traveling whales may encounter members of their own clan or a different clan. Non-relatives and strangers who chance to encounter one another and realize they share a dialect may socialize. If they don't share a dialect, they will avoid contact and socializing. Only in sperm whales and humans do group identities extend so far beyond kin. Sperm whale clans constitute a kind of national or tribal identity at a scale larger than any other non-human.

When something like a vocal dialect or sperm whales' codas functions to differentiate, potentially to segregate or aggregate a group, it's called "symbolic marking." It's still often considered distinctly human, but you and I — and the whales — know better.

A cultural group is a collection of individuals who have learned, from one another, certain ways of doing things. In a culture, Shane tells me, "you are who you are because you're with who you're with. Because of who you're with, you do what you do in the way that you do it." The cultural differences among sperm whale clans include different clans' distinct patterns of movement, diving, hunting, and so on. Each clan has found different answers to the question "How can we live where we live?" And as Shane says, "Clans are how they have institutionalized their answers."

From group to group, the genetics are the same. "What gives clans their sense of identity and their ways of doing things," Shane explains, "is their social learning. Each whale must learn their social traditions. Behavior is what you do. Culture is how you've learned to do it." Social learning is sometimes called "the second inheritance." The first, of course, is our genes, inherited physically from our ancestors. Customs, too, are handed down by elders, but they must be learned. Genes and culture, two inheritances — and both evolve.

Only once, in the Pacific, did scientists see two different clans on the same day. Says Shane, "The self-separation between clans is pretty much absolute." We humans can understand what it feels like to recognize

67

similarities and differences that have nothing to do with genes. In-group and out-group identities depend on what you've learned from your family and your friends, growing up. Raised in a different place among different individuals, you'd be part of a different culture. In that sense, a major aspect of culture is that it is *arbitrary.*

Sperm whales have the only known cultural groupings that exist at transoceanic scale. Everywhere sperm whales have been studied — the Galápagos, Indian Ocean, Gulf of Mexico, Canary Islands, Azores, Sargasso Sea, Mediterranean, Brazil, Hawaii, Mauritius — researchers see attraction within clans, repulsion between. Clan members live united — and thus clans live divided — by their clan identity. Shane emphasizes this: "They live by the difference between 'us' and 'them.' "

In many ways — so do we. This sense of self as an individual among recognized individuals, this layered perception of identity, is considered rare in non-humans. But few other animals have been studied extensively. We know that various whales have it. And bats, too, have similar ways to identify and announce themselves and their local membership. If it's as widespread as the span between whales and bats, could we say, "And probably everything in between?" Probably not. But between whales and bats, it *is* likely that *many* other animals identify themselves and

68

recognize individuals. I expect we will discover more.

Actually, dogs do this implicitly. They know who's who and that we're a unit. When my wife and I are on the beach with our dogs Chula and Jude running loose, they meet other dogs and say hi to other friendly people. Often they'll engage in some play with other doggies or indulge some petting. But if I simply keep walking, they stick together, wait for one another, then come and join me. They never want to go off with other dogs or other people. Our dogs know the difference even between new and familiar dogs. On the beach — this is just an example — Chula will stop to check out a new little white dog, then come running over to me as I've continued walking. After we hit the endpoint of our walk, we turn around and go back up the beach, toward our starting point. From a hundred yards away we see that the little white dog is still there. Chula shows no interest. She knows she has already met this dog. But the moment a tan boxer appears just a few paces from the little white dog, Chula beelines to him. She knows this is a new dog, one she wants to check out. Recently we adopted a seven-month-old Aussie shepherd pup named Cady. On day two we brought Cady plus Chula and Jude to a dog-friendly beach. We decided to unleash all three dogs. Cady stayed nearby and came when called.

And, following numerous mingles with other people's dogs, she immediately resumed traveling with us. After just one day in our household, Cady's sense of belonging was operating.

As we will see later on in much more detail, the fact that culture unites individuals into groups and divides the groups has major consequences for the evolutionary journeys of many species, and for the history of life on Earth.

Hal Whitehead and Luke Rendell, who have studied social learning in whales and dolphins for decades, see remarkable similarities between culture in humans and in other animals. They also see profound differences. But they aren't preoccupied with matching the answers of whales to the questions of humans. They simply write, "Culture, we believe, is a major part of what whales are."

Is that true? Thousands of pages have been spent — and largely wasted — on the question of whether other animals "have" culture. There is nothing real about the question, or the word. It depends entirely on definitions, and there are many definitions of "culture." Too many. Because cultural anthropologists and social scientists are humans who make a living studying humans, in their academic journal articles they come up with definitions like this one: "Culture is behaviors and ideas

that humans acquire as members of society."
If our definition considers only what is
cultural for humans, we can never even *ask*
questions like "Where did the human capacity for culture come from?" or "Do non-
humans have any cultural attributes?" And if
a spacecraft from another world arrived and
little green folks descended its stairway,
would we insist that they don't have culture
simply because they're not human beings? A
definition restricting culture to humans offers
nothing.

Like us, the whales are meat and bones and
blood and nerves; whales are warm-blooded
milk-making child-rearing mammals. Whales
share our air but play, socialize, and live
entirely in the sea. All of that makes whales
what they are. Culture in the form of differing group behaviors and vocal identities
makes whales *who* they are.

Getting to the heart of it, isn't it obvious
that other animals don't have human culture?
Whales have whale culture. Elephants have
elephant culture. The question isn't whether
they have the kind of culture we have. The
question is: What are the cultures of various
species? The deeper questions: Who are we
here with? What are the lifeways upon Earth?
And what, really, is being lost as we wipe the
wild map blank?

FAMILIES:
THREE

This morning through the headphones I hear — faintly — a humpback whale singing in the far distance. He is broadcasting, singing to anyone, to everyone, and to no one in particular.

By contrast, the sperm whales we're searching for are very fussy about who they will speak to. Shane is working to detangle *why* sperm whales have gathered themselves into the tribes called clans and *how* the whales decide who they are, who they're with: who is "in" — and who is to be shunned.

I scan the horizon for any sign of the humpback. Humpback whales' oft-photographed high-flying wing-flippered airborne breaches have made them iconic. They often travel so close along shore that, in their season, we frequently see them blowing and fluking while we take our dogs for a run on the beaches of Long Island, New York.

After giving birth in the tropics, mother humpbacks who've fasted for several months trek back to colder waters and their food. Their little ones follow. For their whole lives to come, those young ones will travel the traditional route they learn from their mother. Eventually *their* little ones will learn it from them. Some humpback whales born on opposite sides of the Pacific — the Philippines and Mexico — trek to the same feeding grounds off Alaska's Aleutian Islands, following routes learned from mothers through the ages. Meanwhile, other Mexican humpbacks and Hawaiian-born humpbacks follow their mothers' annual commute to rich feeding grounds off southeast Alaska and British Columbia. The vastly different destinations learned from mothers constitute a key aspect of their culture.

We normally don't think much about the end points of migration routes. When we consider migrations, we think of them as "instinctive." And for many species, it would be difficult indeed to tease apart what aspects of migrations are learned.

But migrations were not always the same as they are now. Centuries ago, native Hawaiians, keen observers of the sea and its creatures, made no mention of humpback whales. This is strange, because nowadays if you look seaward from almost any beach in Hawaii for a little while during winter, you'll likely see one or more humpback whales erupting out of the ocean, detonating explosions. Whales are hard to miss — unless they're not there.

So where were they? It appears that humpback whales began to concentrate in Hawaiian waters only about two hundred years ago. And since the 1970s, their density has risen dramatically, seemingly faster than reproduction would account for. How could such a thing happen?

When the nineteenth-century whaling ship captain Charles Scammon discovered the Mexican lagoons where gray whales congregated to mate and give birth, he almost completely exterminated them. Perhaps a few humpback whales, subjected to such havoc, somehow decided to avoid a breeding area turned fatal, and discovered Hawaii. Their songs — potentially conveying something of peacefulness by, say, their mere completeness — may have been heard by distant whales, who veered toward them as though investigating rumors of some promised land, and found it good. At any rate, it's pretty certain that the whales weren't there in the 1800s. They're there now by the thousands.

No one knows for sure whether humpback whales went to Hawaii to escape their holocaust, but if they didn't go for that reason, their timing was impeccable. This singular diaspora might be the most momentous and successful non-human cultural shift during the recent history of life on Earth.

A mother-to-child process of route learning

likely occurs in all whales. Researchers conclude that beluga whales travel nearly four thousand miles a year along ancestral migration routes by use of "culturally transmitted behavior." Right whales *used* to frequent the waters off New York's Long Island, Canada's Grand Banks and Gulf of St. Lawrence, and the coast of southern Greenland. But whale hunters totally wiped out the right whales of those regions. Remember: whales go where their mothers took them. The remaining right whales of the North Atlantic — only a couple hundred — are keyed in to the southern Bay of Fundy. Unfortunately, food isn't always plentiful there. Right whales have lost elements of their migratory culture that brought them to different foraging regions. Recently, researchers tracked one right whale mother who took her young one through several different feeding areas. She was probably searching for enough food to sustain them. In this there is hope. But it will take more than one hopeful whale.

Change is the only constant, true enough. Change that comes too rapidly, however, brings the end of adaptation, the end of the line. That is the message whispered to us by many extinct species who'd thrived but could not cope with change that hit too fast, bit too hard.

The fact that a humpback whale might sing

— like the humpback singing this morning — was entirely unknown to humans until the 1950s. U.S. military personnel who'd begun listening for Russian submarines were astonished to realize that the strange sounds they were hearing were coming from *whales.* Word got to whale scientists Roger Payne and Scott McVay.

Payne's 1970 vinyl record of humpback songs became an instant sensation. When he and McVay published "Songs of Humpback Whales" in *Science* in 1971, the journal's cover featured a visual representation of the structure of a song. Their first paragraph noted, "Humpback whales produce a series of beautiful and varied sounds for a period of 7 to 30 minutes and then repeat the same series with considerable precision. . . . The function of the songs is unknown."

The song contains elements that make up themes that the whales repeat in a specific order. Payne told me, "The songs of humpback whales employ rhyme — and why not? Humans have been using rhyme at least since Homer and probably long before. It's a way of remembering." A male humpback will usually complete the song, then repeat it numerous times, singing for hours on end. Payne tells me that singing humpback whales normally breathe when the song appears to reach its end. Every now and then they breathe in mid-song, but they don't interrupt their singing. "They tuck their breath in, and let the song continue to flow."

We know now that the male humpback whale's strange and haunting singing is a changeable cultural aspect of his species. Each year, all adult male humpbacks within each ocean sing the same song. But in each ocean, the song is different from the song being sung in other oceans. There's a Pacific song, an Atlantic song, and so on. And each year, the song of each ocean changes. The new songs spread wavelike, a slow-moving fad crossing blue infinities whale to whale, all the whales adopting the same changed elements of the song. How it will change, how much will change, and how rapidly, humans cannot predict. Somehow together, strangely, the whales create a new song. There's a lovely metaphor for us in that. When the song of Hawaiian humpbacks and Mexico's Socorro Island humpbacks changed simultaneously despite a separation spanning nearly three thousand miles of ocean, researcher Ellen Garland and her colleagues called this pattern "unparalleled in any other nonhuman animals . . . culturally driven change at a vast scale."

Planet Earth constantly thrums with messages being sent and received by living things. Life is vibrant, and it generates good vibrations throughout the air, the sea, and the ground. But whale sounds seem particularly enchanted. Roger Payne wrote of the first time he heard a humpback whale singing: "Normally you don't hear the size of the ocean . . . but I heard it that night. . . . That's what whales do; they give the ocean its voice, and the voice they give is ethereal and unearthly." Payne later told me, "The reaction of some people to hearing whales sing is to burst into tears; I've seen that a lot."

How the songs function among whales remains unknown. Females do not approach singers; nor do other males. The function of humpback whale song among humans is easier to describe: their songs tangled human emotions, creating a momentous turning point in the human relationship with life on the planet — and beyond. Humpbacks sing the song of their kind and their culture, but it resonates in humans who hear it. In 1979, *National Geographic* magazine inserted a disk of humpback songs into ten million copies of its flagship magazine, the largest-ever printing of any recording. Not only was Payne's life changed; his recordings changed whales and humanity.

The first thing the recordings did was to save the whales from total annihilation. Propelled largely by the beauty humans perceived in these recordings, the "Save the Whales" movement hit full stride. Humans learned that whales are not things but, rather, neighbors living with us in the world. And so astonished were we by this realization that whales went from being ingredients of margarine in the 1960s to spiritual icons of the 1970s' emerging environmental movement.

After the first recordings, musicians such as Paul Winter, Judy Collins, and David Rothenberg began making music with humpback songs. As the volume of the music came up, the blasts of the harpoon cannons receded. Within a few years, whale hunting was largely ended.

So deeply did the whales' music reach us that a recording of humpback whale singing is among the few sounds included aboard the *Voyager* spacecraft. Humanity's calling card to the galaxy has taken humpback song beyond our solar system. It's a message in a bottle — humankind hoping, perhaps, that an alien life-form of great and cultured intellect will understand. But the whales' message is simple, and we ourselves should be able to understand it: "We, the living, celebrate being alive." The song culture of humpback whales changed our interspecies culture. And why? Simply this: we briefly directed our attention to something beautiful on Earth. For a moment — we listened. The whales continue calling us, asking, in effect, "Can you hear me now?"

After some interval that I am not closely tracking, our boat is undulating across the massage of long swells. We transit in and out of the company of flying fish. Of terns. The sea, glittering, rolls like a carpet of short blue flames. We travel in small ecstatic sparks of time.

Out of the chop, from up ahead, comes a large group of dolphins. Short snouts, small dorsals, pastel-pinkish bellies and throats. Medium-sized; a bit longer than a human. I've never seen these.

Shane identifies them: Fraser's dolphins, a species unknown to humans until the 1970s. We slow to a near standstill while we document their position and their numbers. Several groups, about eighty individuals. Many small babies. They evince no fear, exhibit no evasion. They rise together and dive together, coming into view through the surface as a bunch and going out of view through the surface as a bunch, so that the visual effect from our side of the sea is as if a group of dolphins are riding a Ferris wheel as they travel. Their food requirements could be calculated. What is incalculable: the nerve-and-muscle journey that life undertook through deep time to form and power the speed we are witnessing, and our capacity — theirs and ours — to sense beauty and grace in all the exuberant leaping, gravity-defying somersaults, sun-sparkled aerial spins, and fun-enriched lives.

The hydrophone goes in just so we can

listen to them. The headphones bring our minds into a more fluid universe. The dolphins' squeals and whistles seem liquidy, less electronic-sounding than those of the *Pseudorca* whales we'd heard the other day.

Information is code. Communicating information does not require conscious intent; computers do it. In the living world, plants — though they may lack felt experience and intentionality — convey lots of information. Flowers are code for the presence of nectar and pollen. Flowers are the plants' ad campaign for getting hungry bees, bugs, birds, and bats to pollinate them. What they're advertising is: Come visit our flowers; we're offering nectar and pollen (the plants, appealing solely to the needs of their pollinator clients, don't explicitly mention that all they really want is sex). Bright fruit later communicates the ripeness of sweet nourishment for fruit-eating animals as a way of getting their seeds spread. Plants also generate much chemical communication to other plants and to insects that eat other insects, to help alleviate attacks from leaf-eating bugs.

Animals produce signals coded in sounds, scents, songs, dances, rituals, and language. Humans use language so much that it swamps our own ability to recognize subtle and not-subtle nonverbal signals that we ourselves continually display and respond to. Many other animals, too, use meaning-bearing gestures. Some have a small vocabulary of words and even simple syntax. Many species have dialects. Sperm whales have their strangely beautiful codas. The world is awash in layers and waves of communication.

The farther from the ever-more-distant green slopes we travel, the bluer the sea.

And now, another dark fin. A small whale — dolphin-sized — alone. Barely moving. Nondescript. I'm perplexed.

Shane, at a glimpse, says, "Could be a beaked whale. But not Blainville's or Cuvier's. Maybe Gervais' or True's —"

A downward-looking royal tern comes over, cuts an abrupt semicircle, plunges, and lifts from the sea in its orange bill a needlefish seized by the head, its body writhing and its long jaw agape.

"— except that baby beaked whales dive with their parents."

So is this a distressed baby?

The hydrophone detects no distress callings.

"Did you see how the blowhole is forward-facing and large?" Shane asks me.

I did not. I'm doing little more than staring. I'm ignorant even of what distinguishing features to look for.

"That dorsal —. It all makes me think it could be *Kogia*."

Far smaller and rather unlike actual sperm

whales, two species that were long thought to be one — now called pygmy and dwarf sperm whales — are classified in the genus *Kogia*. They're nearly impossible to distinguish from one another at the surface. I once saw a pygmy sperm whale who'd washed up alive near my home on Long Island. Less than ten feet long, the stricken creature seemed a strange contrivance, like an early mocked-up prototype of a sperm whale, with a smaller head and a smaller mouth, and a much smaller body. Yet there it was, the result of millions of years of becoming a distinct being. And here it is. And there we leave it.

I am struck by the oddness of certain names. Pygmy. Dwarf. False killer. Names of diminishment and dismissal. Names reflecting some confusion. They are real species named as if not the real thing. Of the many torments humans have inflicted on whales, not least is the nomenclature they must drag around the seas like little dories of indignity. Hunters branded whales with some awful names, and scientists have perpetuated them. The reluctance of marine mammal scientists to update names stands in sharp contrast to the practice of bird scientists; their opposite but equally irksome tendency is to continually shift Latin and common names of birds while indulging in frequent lumping and splitting of closely related species. (For example, the bird formerly called the common gallinule because it is in the genus *Galli-nula* was in the 1980s lumped with the European form and renamed "common moorhen" — though they don't live on moors and any fool could tell you that they're not all hens — and then in 2011 they again became the "common gallinule." Nothing about the bird itself changed before or after all the name-calling.) Whale people, on the other hand, stubbornly stick with names and designations, no matter how inane or anti-quated.

For instance, the creature designated "fin whale" is but one of many fin-bearing species. The "humpback" doesn't have a particularly humped back, but it does have by far the longest pectoral flippers of any whale. And at times a humpback uses those wing-flippers to propel and maneuver, as penguins use their flipper-wings. The humpback *could* be called the long-winged whale. In fact, its Latin name means "long-winged New Englander," which comes close to making sense except that humpbacks live everywhere: in the Atlantic, Pacific, and Indian Oceans, into the Arctic, and around Antarctica. So — New Englander?

The right whale was the "right" one to kill because its corpse floated. That was in the days before engines and harpoon guns made any whale "right" to kill, wrong though that was. May we please now have a better name for this still-beleaguered creature? (Could renaming it the "let-it-be whale" be worse than continuing to call it the "right whale" to kill?) And saving the best for last, we have the sperm whale, named, yes, because part of the amazing head is filled with a substance that reminded seamen of semen. They were, of course, totally ignorant of the substance's sound-production function. Another name for this whale is "cachalot." In Portuguese, *cachola* can refer to the head. One wonders whether, through accident or intentional wordplay, English-speaking whalers corrupted this foreign reference so that it sounded like "catch-a-lot." We have a "right whale" to kill and a sperm whale to "catch-a-lot." The names reveal more about us than about them.

FAMILIES:
FOUR

The whales we seek, free in their liberty and unbridled in their pursuits, inhabit a world as unbounded as any. The water floating us is fully three miles thick. Dark even at the surface. This place, so thoroughly inhabited by unfamiliar creatures, feels inaccessible to humans. Because it is.

Schools of fish that might be Spanish mackerel suddenly begin tearing up the surface, and flying fish fleeing the attack launch into the breeze on cellophaney wings, as if created from the sea spume off our bow. Here in Wonderland, fish that fly waft long distances, merely gliding. When they nearly touch in, they buzz the water with their tail's lower lobe, relaunching, extending their flight paths. Their bellies mirror light shards bouncing across the wrinkled sea. That hides them from fishes. Their dragonfly wings are the color of air, making them nearly invisible when aloft. Nonetheless, flying fish find no assurance in any dimension. In the water, death lurks. In the air, death waits.

Swirls on the surface mark mackerels' misses where flying fish have launched into the wind. A swoop from above and a flying fish drops into the sea. The surface turmoil has swiftly drawn hungry frigates and several streaking boobies. Frigates snatch the flying fish from the air, and the boobies can also plunge through the surface as the fish ditch. Often enough for the fish, they miss. Often enough for the birds, they seize the day.

Only by going through the looking glass can the flying fish achieve some reprieve from the constant imminence of well-aimed death. Everywhere, eyes search for them. Below, fish pursue them and the sonar of dolphins pings them. Above the surface, birds streak after them. Flying fish fall prey in uncountable millions. Yet only by thriving in greater millions — as they do throughout warm seas — can they support their enemies. All success — the flying fishes', the birds', everyone's — is temporary, but temporary success is everything.

I would linger, absorbing the life-and-death beauty and eruptive confusion of frenzying fishes and birds. But my mammal-seeking companions have no time to tarry. They've seen it. I watch the diving birds and water-shredding fishes recede in our streaming wake.

One flying fish leaps away from our hull,

catching my eye, gliding on and on. The distance this one glides leaves me incredulous. I'm thinking three hundred yards. Shane is watching it, too. And when the fish splashes in, I ask how far he estimates. "At least two hundred meters," he says. "Could be more," he allows. "That was far."

On this day's eleventh drop of the hydrophone, something is making interesting whistles. Distant. Hard to hear through the sloshing of the sea surface. And wait — something else —. Quick little clicks.

Half a mile away, one small whale pokes a head up, having a look around. We venture closer. About thirty dark-bodied creatures. Small, less than ten feet. Heads wide.

Another first for me: melon-headed whales, very much on the hunt, chasing small fish. The fish leaping for their lives have a dark curving bar running from their back down through their tail's lower fork. They are called bar jacks. We're soon joined by fluttering dark noddies, frantically dipping bridled terns and royal terns, and more frigatebirds. Now it's a frenzy.

Not far away, a single dolphin begins a series of astonishingly high leaps. Again and again the dolphin flies, the peak of its arc well above the horizon. Shane speculates that it's a Fraser's dolphin because "that's who melon-headed whales are often with."

And as if the leaping dolphin was calling friends, about thirty Fraser's dolphins come suddenly bounding in out of nowhere, their pink-bellied bodies streaking among the dark melon-headed whales. Playing? Competing for the same food? Using the whales in some way? Each fractal of fact shatters into shards of questions.

We are headed home from this mammal-filled day when a large group of dolphins and their numerous babies swarms us. They are yet another species — pantropical spotted dolphins, known in the biz as "pantrops." They want to ride the wave created by the bow of our moving boat, and they don't ask permission. This is what freedom looks like. I watch them in the clear water. They stream along, pumping and pacing. They leap and plunge and turn sideways to look up at us looking down. Their bodies are dotted individually, some with many spots, some with very few. They precisely exhale streaks of silvering bubbles before snatching a sudden gasp of breath and pursing their blowhole shut and surging under, all at full speed inside of a second. The water here is sprinkled with drifting yellow seaweed, and some of the dolphins playfully snatch pieces of it on a flipper as they stream along. Effortlessly. Miraculously.

■ ■ ■ ■

For a long time, a silly debate raged about whether animals (including humans) live entirely by instinct *or* by learning. The debate was called "nature versus nurture." Genetically fixed instincts were "nature." Learning and culture represented "nurture." Some people believed that all animals, including humans, are born as tabulae rasae, literally "blank slates" having no instincts, and that all behavior is learned. Others believed that all behavior is instinctive. Not very realistic, or even observant. Both nature and nurture are in play; both interact. Genes can deliver results, but they don't always dictate *which* results. The physical and behavioral expression of genes can be tweaked by the environment; this is referred to as "epigenetics." Humans are genetically enabled to acquire any human language. But we must still *learn* a language. Genes facilitate the learning, but they do not determine whether we will speak Russian. Genes also delineate what cannot be learned. Whales can learn, but they cannot learn French. Humans can sing, but not like a humpback. Human genes facilitate human culture. Genes facilitate a *social* creature's ability to learn socially. Human genes do not enable us to hunt squid using sonar or, like elephants, send rumbles across the ground to alert family members to danger miles away. Nature (genes) affects the kinds of nurturing (learning and culture) that are possible. Genes determine what *can* be learned, what we *might* do. Culture determines what *is* learned, *how* we do things. You might say that for many matters of knowledge and skill, genes let us know there is a question, without giving us the answers. That's because there are many answers to some questions, and from place to place the best answers can vary. The varied answers are culture.

The philosopher Immanuel Kant asserted, "Man is the only being who needs education." But if he'd said, "*not* the only being," he'd have been correct. Many creatures *depend* on learning. That's why so many animals are capable of it. A kitten instinctively chases. But the cat must learn to hunt effectively. Kittens who get to watch their mothers quickly become better hunters than kittens who have to figure out, on their own, what their claws and teeth and curiosity are for. A lot of learning travels socially from parents to offspring or from a group's elders.

The power in learning, whether solo or social, is that you get information that isn't in your born brain. Learning allows an individual to grasp great hunks of information directly from the world. You find food in a certain place. Next time you're hungry, you'll return; you've learned something valuable that will help you survive.

Social learning is special. Social learning gives you information stored in the brains of other individuals. You're *born* with genes from just two parents; you can *learn* what whole generations have figured out. Social learning can create and spread changes in behaviors and group customs much faster than can evolution that relies on DNA mutations, the winnowing of survivors, and the slow spread of changing genes.

(Let's pause here for one parenthetical paragraph to clarify that physical evolution is a slow and incremental process. Evolution isn't the sudden appearance of new species, nor of instant large mutations. Radical mutations are usually fatal. Evolutionary changes are generally slow, advantageous riffs on the average, such as longer-than-average legs or a slightly shorter and harder beak that facilitates more efficient access to a wider array of seeds. For example, the evolution of whales from the land mammals didn't happen because a litter of land grazers were born with flippers instead of legs. That bridge-too-far mutation wouldn't work. Rather, the first step was for a wetland-dwelling population to develop the advantage of webbing between their toes, like a Labrador retriever's, leading to a more aquatic life benefiting from having webbed feet like an otter's, leading to flipper feet like a sea lion's, then to flippers like a seal's that remain like flexible hands in mittens with nails that can scratch an itch, and eventually to the stiff fin flippers of a whale. Elapsed time: several million years. But even a whale's pectoral fins still retain all the same bones as your shoulder, arm, and fingers. Any survival advantage of a certain slight difference translates to a few more offspring than the average, thereby increasing the frequencies of certain genes in a population. In fact, evolution can be defined as "a change in gene frequencies." Sometimes the whole species evolves in this way and changes greatly over time. Sometimes the changes happen only in one population or one group, and diverging differences eventually become great enough to inhibit breeding between the groups. The reproductive estrangement of two populations inaugurates a new species.)

With *social* learning, an individual who is new and naïve in the corridors of the world gets the keys to the doors and drawers and cabinets of collective knowledge. You get skills tailored to what you happen to need, where you happen to be, as an inheritance from the whole community. It's a great leap over learning solo by trial and error, that fraught process of acquiring skills at the cost of time, chance, and, sometimes, mortal risk.

Social learning is huge, because it means that a dolphin or an elephant, a parrot or chimpanzee or lion, can tap into collective skills and wisdom that accrued slowly over centuries. For a young whale: Where in miles and miles and miles of ocean should I look for food? For a young elephant: Where is drinking water when everything I know has dried up? For a young chimpanzee: Now that the fruit is gone, what do I eat? For a young elk: As everything begins freezing solid, where should I go? For a young wolf: How might we hunt and eat this creature that weighs ten times what I weigh? These are all learned skills. For many creatures, they are skills learned from experienced elders.

Many things that you crucially need to know, you would never learn on your own. We learn from others who already know. From one another we become who we are. The things we learn socially answer the question "What do we do to live where we live?" This is true for us humans, and it is also true in surprising ways for a world-spanning array of other species.

So we become equipped with knowledge in at least three ways. There's genetic inheritance (instinct), trial and error (individual learning), and social learning (customs, traditions, culture). The things we learn socially give us more than just skills. They also give us group identity, conformity, unity — and divisions. Culture.

■ ■ ■ ■

One definition of culture that is pretty good is: "the way we do things." Behavior is what we do; how we do it — is culture. Reach for the leash or your car keys and your dog immediately gets excited by the prospect that you'll be going somewhere together. Shared culture.

But one big thing is missing from that definition: to have culture, someone must do something that is *not* the way we do things. We live in an automobile culture, but only because an innovator invented an automobile. We listen to rock music, but one person electrified the age-old guitar.

Ironically, culture — a process of learning and conformity — depends on individuals who don't entirely conform to the way we do things. Culture depends both on doing what you've seen done — *and* on someone, at some point, doing what no one has ever seen done. To encompass both the ways we do things *and* innovation, our definition of culture can be this: culture is information and behavior that *flows* socially and can be learned, retained, and shared.

This may be why human cultural generators — inventors, designers, innovating artists — are often reclusive, quirky, and introverted. At any rate, culture depends crucially on

95

crowds of conformists *and* the rare innovator. Without some original innovator — some untaught learner, some unschooled teacher — there is *no* knowledge, skill, or tradition that can get shared; there is no culture to copy and conform to. Innovation is to culture what mutation is to genes; it's the only way to make any progress, the root of all change.

So even though a baby whale follows her mother to one of the species' traditional foraging spots, the only way such a tradition can start is that, every now and then, someone has to break with tradition and go a new way.

The chances for a human to see, perceive, and record a pioneering animal innovator are few. And yet, in 1980, one of the humpback whales off New England started whacking the water surface with his fluke several times before a dive if he was feeding on a particular kind of fish called sand lance. He apparently did this to create a loud disturbance that frightened these little fish, making them clump closer together. Tight crowding is good defense for small fish under attack by many bigger fish. Safety in numbers. But grouping up makes little fish more vulnerable to humpback whales, whose cavernous open mouths can engulf whole schools. Over ten years, this slapping technique spread to half of the whale population, as one young whale after another copied a close associate who was already schooled and skilled. (Older whales weren't copying.)

In some regions, sperm whales have learned to take hooked fish off longlines that are miles long and dangle hundreds of baited hooks. This is a recent tradition that, for the first time in history, puts some sperm whale groups in a position to turn the tables on human interests rather than always getting the bad end of the harpoon, of depletion of their food, and so on. (But it's not a strategy I recommend; it brings the rifles out of the wheelhouse.)

Humpback whales sometimes decide to forage in groups, using a coordinated strategy called bubble-netting. One whale, or a few, had to have innovated the blowing of circular "bubble nets" to corral schools of fish and other targeted food. If you've ever been lucky enough to catch them doing this, you've seen several whales — maybe six or eight — dive together. You've watched whale-sized rings of bubbles begin coming to the surface as the deep-circling whales blow a rising curtain of air that scares and confuses the fish into a concentrated center. Perhaps, like me, you've stood in amazement as the whales coming up through the middle of their big bubble rings suddenly burst the surface, lunging up, mouths open, with dozens of little fish leaping out of those fatal swimming pools in the final moments before the jaws of death close on them.

Ecologist Ari Friedlaender has studied bubble-netting in great detail and tells me that in Alaska, groups with more than a dozen whales bubble-net for fast-moving herring, while off New England, where the prey fish are often slower-moving sand lances, the bubble-netting parties are smaller. The Alaskan groups tend to be more stable, sticking together. New Englanders seem more opportunistic. Perhaps most interestingly, whales specialize. For instance, it appears that not all of the whales in the party create the bubble curtains, but a whale who blows bubbles in one sortie is a whale who'll blow bubbles in the next. The whales carry a quiver of cultural choices. They draw upon what they need to fit the occasion.

A while ago, some researchers wanted to see how long it takes a killer whale (orca) to imitate. In this experiment, researchers arranged for a human trainer to work with three captive-born killer whales who were show performers. (Nowadays captive breeding and shows are on the way out, but for a few decades a lot got learned about *why* these beings are not appropriate subjects of our amusement.) The trainer asked one whale to do something that whale had previously learned to do on cue, such as slapping a fin. Then trainers pointed and used eye contact to ask another whale — who'd never done it — to imitate the action. In just one or two sessions, the researchers reported, "All three subjects copied correctly 100% of the untrained behaviors." (If you've ever tried to teach a dog to roll over, you know it takes many steps, rewards, and practice.) During that study, a two-month-old infant, with no prompting and no reward, instantly imitated three tricks her mother performed. A capacity for social learning can spread one individual's specialty into the customs and culture of many.

Sperm whales coming up through the ocean toward the blanketing air will often announce their individual identity and their group membership. Using their codas and their dialects, they show and declare, "This is who I am. This is with whom I belong."

Pushing continually through the seas, sperm whales have only the long deep swells, the cold dark waters, the wide wild skies — and each other. Knowing who they are because they've learned whom they're with, together they face their foes, find their food, and confront their fears.

When a family of whales is "chatting," as Shane says, there's a lot going on. What's known: using learned codas, sperm whales can announce themselves as individuals and indicate their family and their clan membership. Each whale makes a coda about every five seconds. You hear lots of overlapping.

"It's not rude to talk over each other if you're a sperm whale," Shane says. Often a particular coda starts an exchange, as if to say, "Can I have your attention." The first codas of a down dive often come in a cadence of five regularly spaced clicks. Then the whale might switch to a different coda.

Coda dialects serve both as social lubricant and, equally importantly, as social barrier. As boundaries of human societal groups can be marked by language differences, sperm whale society boundaries are reflected by differences in group codas.

The way sperm whales use their codas is complex. So let's start simple. Each clan favors various codas. Worldwide, scientists have identified more than eighty coda types. Caribbean sperm whales make a total of twenty-three different codas. Some families work with nearly all of the known codas. Some omit many. Here in the Caribbean, there are just two clans. In each clan's overall vocabulary, there is *one* coda that is used *only* by the members of that clan.

Those clan-specific codas are a little bit like needing to enter your access code and hit # before you can join the conversation. Remember, families of a clan talk to each other, but clans don't talk to other clans. Almost two dozen families claim membership in one of the clans that move through Dominica's waters. The second clan is made up of just a handful of family units, who show up occasionally. Other Atlantic clans live in the Gulf of Mexico, the Sargasso Sea, around the Azores, and elsewhere. Their overall ranges sometimes overlap. But, as mentioned, different clans apparently won't socialize.

In the whales Shane has studied for over a decade it turns out that a clan's most commonly used coda is, in fact, the one that identifies the clan. The second most common identifies an individual, like saying your name is Bonnie. But to identify your family, you have to say you're Bonnie Thompson. That's why the third most common coda identifies the family.

So now let's speak a little sperm whale. The larger of the two clans here makes a five-click coda that has never been heard from any other sperm whales in the world. It goes, "One, two, cha-cha-chá." Shane calls it "the $1+1+3$." It's a marker that says, "I'm from the Eastern Caribbean Clan. Are you?" Whales who make the $1+1+3$ coda spend time together. Those who don't, don't.

The other clan's identifying coda is a long, slow, and, Shane says, "more deliberate" sounding "One, two, three, four, five." Those whales draw this coda out to almost three times the duration the other clan takes for its quick "One, two, cha-cha-chá." So when whales here are making various codas, you may hear *either* "One, two, cha-cha-chá," *or* the long, slow five-click announcement. "Once we hear either of those, we know which clan it is," Shane says.

"All the whales in all the families of a clan teach their young to make their codas in this explicit way," explains Shane. Since it was first noticed thirty years ago, it's been the same.

The second most commonly spoken coda is five clicks performed rapidly, which Shane calls the 5-R. "And *that's* the coda," Shane says, "that lets us reliably identify individuals within their family. That lets us hear two whales and say, 'That's Fingers, and that other one is Pinchy.' " How? "In a family, each whale has a minor but consistent difference in the timing between the clicks," Shane says. "Pinchy might make the first click slightly longer, and Fingers might make the last click slightly shorter, for instance." You might record a 5-R coda from two different whales, then put them on a graph of time and see that one whale made five regularly spaced clicks a bit slower and one made them faster. It's the same coda type, but if you compare whales, none of the clicks line up exactly in time. It's possible that whales also recognize individuals just by the sound, as we recognize people we know just by voice, the sound of their "Hello." If a stranger says hello, well, you don't know who they are.

"After spending a lot of time with the Group of Seven," Shane says, "we realized that Fingers sounded different from the other moms. She made one particular coda far more frequently than any of the other whales. She tends to start the vocal exchanges, tends to decide to dive first, and she's the most socially gregarious. We think it's a signal that she's the leader."

However the whales perceive it, within families all the whales know each other as individuals, as we know family members in our own homes. That's not to say their relationships feel like ours. But their recognition is as complete and immediate as ours, or as dogs greeting their own people. There's no mistaking. The difference is: the whales are never far apart. Swimming thousands of miles, they never lose contact; they can always hear each other calling. Always. That's closeness.

The third most common coda in this main clan — the one used to identify the family or "unit" — is always four clicks. But one family makes a certain *pattern* of four clicks; another family makes a different four-click pattern. Unit N makes a fast 1+3. Unit V makes a longer 1+3. Unit F's 1+3 is in between. Unit A makes a relatively long 4-Regular. Each pattern is consistent in its family-to-family difference. By this coda, each family within a clan is recognizable.

"We usually hear it only when there are *other* families around," explains Shane.

So — translated to English — the whales can say, "I belong to the Eastern Caribbean Clan. I am Pinchy of Unit F."

"That is self-recognition and self-identity. That is a community of families and individuals. These whales," Shane declares, "are cultural beings."

After explaining it to me simply, Shane starts giving me the complexities. As mentioned, each clan here in the Caribbean has *one* identifying coda used by *no* other clan. But that's *not* the case in most of the world. Worldwide, sperm whales' use of codas gets very complex indeed.

It took me a while — and several attempts by Shane — to understand what first seemed pretty confusing. I think I got it, finally. It must be easier for the whales. Anyway, here we go:

In most of the world, the dialect differences between clans are a matter of how frequently certain clans use certain codas. Imagine that one musical group likes a few particular chords and uses them a lot. Another group prefers several different chords. Both groups use all the same chords, but each group uses certain chords more frequently than does the other group. And this helps give each musical group its distinctive, recognizable sound.

Sperm whale dialects are a bit like that. The codas each clan *tends* to make more often give the clan what Shane calls a "thematic" distinctiveness.

For instance, all three clans off the Galápagos Islands make the 5-Regular coda of five evenly spaced clicks. The Regular Clan makes it a lot. But the Plus-one and the Short Clans seldom do. The Plus-one Clan most often adds a long beat before their final click in any of its codas, so whether they make four, five, or six clicks, they pause for a moment before making the last "click." The Short Clan makes short 3- and 4-click codas more than other codas. And so on for several other clans.

We've come out again this morning as we have each day the weather has let us, and not long into our listening we picked up sperm whales. Now most of the whales we've heard are at the surface, widely spread out and rather quiet.

Their coordinated surfacing is no coincidence. They know where they are relative to each other, and who's where. To themselves, the whales feel very much together. To us, they reveal so little. Blowhole to dorsal fin along the ridge of their dark gray backs, we can see. The smooth, vast head, so bulked that it actually bulges from the line of the back; the wrinkled skin of the body, so well

105

tailored to the wrinkling skin of the sea. But who is touching whom under the table, and whether they are gazing eye to eye, they alone know.

And now their leisure is over. There is a living to be made. Of the nearby whales, the largest hunches her puckered back and begins flowing into her dive, lifting her tail flukes to the sky, letting the massive weight of her aft end pile-drive her into the deep. A second whale raises the wide black flag of her flukes and flows so gracefully into her deep dive that she seems as timeless as the rolling ocean itself. "In no living thing are the lines of beauty more exquisitely defined," wrote Herman Melville, "than in the crescentic borders of these flukes." And soon the other distant dark backs also lift, and the flukes rise into the air, and all the whales pour themselves back into the ocean.

The swells sweep closed the broken sea, hiding all trace of the whales' visits and departures. Shane looks at the photos we've just gotten of all their flukes. It's the J Family, most of whose names are inspired by the play *Oedipus Rex*. "Okay," he says. "So of those first two adults, the mom is Sophocles. She's the mother of the young one, who is Jonah — who, despite her name, is actually a *female*. When most of the family was down hunting, Laius, who has two nicks and a wide scoop missing, was the one babysitting

106

Jonah." Shane lingers on the frame and breathes, almost to himself, "Jonah needs to make it."

Their commute to the food-bearing vein of the ocean takes about ten minutes. And when they arrive, they inhabit a realm requiring all their special gifts and superpowers. The diving record for a tagged sperm whale so far is 1.2 kilometers (three-quarters of a mile). A Cuvier's beaked whale, also tagged, went to more than 2 kilometers. However, a fresh bottom-dwelling shark was found in the stomach of a large male sperm whale that was killed in water 3,190 meters deep — strong circumstantial evidence that the whale may have been foraging at a depth of almost two miles. These are the limits of what mammals that breathe air like you and me can do. These mammals are not like you and me.

This is the pattern of their typical day. They'll be down for forty to sixty minutes, eating where sunlight cannot pierce, swimming a couple of miles, then rising to the distant surface.

So we eat our own lunch, vegetarian fare whose health pretenses are negated by our team's weakness for sugar cookies.

The sky is a light hazy blue. The water, dark. The breeze fresh. Having noted their compass orientation as they dived, we will motor slowly to somewhere within plausible proximity for their next resurfacing, an hour or so from now.

Sometimes they surface in synchrony: *Pfff, pfff, pfff, pfff.* Sometimes they unite at the surface, getting physical, rolling with each other, rubbing, caressing with their short front flippers, running their mouths along each other, producing lots of codas and buzzing each other with their sonar. Sometimes they suckle each other's babes. Sometimes when descending in a deep dive, they're together, touching each other. What does it say to us, of them, that these beings seek physical contact, that they love to touch?

FAMILIES:
FIVE

It is a matter of great astonishment that the consideration of the habits of so interesting, and in a commercial point of view of so important an animal, should have been so entirely neglected, or should have excited so little curiosity.
— Thomas Beale, 1839

Nothing could quite prepare people for sperm whales' strange ability to hunt in the dark using sound. And nothing quite did.

Lacking the hydrophones we have, early whalers never heard these whales make any noise. "The sperm whale," wrote Beale in 1839, "is one of the most noiseless of marine animals."

But whalers did understand that the whales were listening. Carefully. Though they believed them silent, they realized that sperm whales had a way of instant-messaging one another through large distances of seawater.

Beale informs us that it is "necessary for the whaler to be extremely cautious in his mode of approaching them . . . for they have some mode of communication one to another, through a whole school, in an incredibly short space of time."

Herman Melville echoed the idea that sperm whales possess "some instinct" by which to rally large groups when confronted with danger. In his novel, Melville sets three whaleboats toward a pod and writes, "No sooner did the herd, by some presumed wonderful instinct of the Sperm Whale, become notified of the three keels that were after them, — though as yet a mile in their rear, — than they rallied again, and forming in close ranks and battalions, so that their spouts all looked like flashing lines of stacked bayonets, moved on with redoubled velocity."

Jacques Cousteau speculated in *The Silent World* (apparently without realizing the irony) that dolphins might use sound to navigate. By the early 1960s, dolphin sonar — formally "echolocation" — was finally considered confirmed. The pioneering researcher Ken Norris showed that tame captive dolphins wearing suction-cup blindfolds effortlessly retrieved toys, snatched fish, and negotiated around clear panels.

The way it works is, the creature generates vibrational energy — sound — and then detects and deciphers echoes of that sound bouncing off objects. It's a simple concept. But it's mind-boggling when you consider that sonar-enabled animals can simultaneously generate sound, analyze echoes, pursue at high speed, and capture fast and evasive prey with their mouths.

Land animals, of course, evolved sophisticated, specialized structures for breathing and managing flows of air. Whales, seals, and sea lions took those land-evolved sound-making air boxes into the sea. Controlled air is excellent for producing sound underwater. Some fish use the gas in their swim bladders to grunt calls to one another (fish in the "drum" family are named for their tom-tom calls).

Around thirty-five million years ago, an ancestral whale began using sound and echoes in hunting in the sea. (Bats' sonar evolved separately.) Rudimentary sonar must have helped during nighttime hunting of small fishes whose vision and evasion skills are keen in daylight. It must have greatly surprised the multitudes of squid that come from great depths to the sea surface during darkness. Then, with the system refined to master nighttime near-surface hunting, whales could evolve to follow squid into the dark depths where they hide while the sun shines. Shallow and deep, sonar proved an enormous advantage. Its bearers diversified. And so we have the dolphins and other sonar-wielding toothed whales we know today.

Squid often hide very deep. To get at them, evolution eventually equipped one creature

111

— the sperm whale — with Earth's most extreme biological sonar. So effective is Leviathan's ability to hunt in deep darkness using sound that Beale wrote that a sperm whale can remain well fed even if blind. That extreme sound system makes Leviathan the planet's strangest-looking supersized predator. Behold: the gigantic head. Weighing as much as ten tons, making up a third of the whale's body length — in larger males it can span *twenty feet* from nostril to eye — the sperm whale's head is by far the most beautifully weird sonic generator in the living world. The sperm whale is the whale of literature and of bathtub toys, the whale of caricature and of accoutrement. So that strange head is oddly familiar. Ask a child to draw a whale, and you'll usually recognize that head. So bizarre is it, its function remained a matter of speculation for decades. There were plenty of guesses. Maybe it was for buoyancy regulation. Maybe it had evolved mainly as "a product of specialization for male-male aggression . . . to injure an opponent." That last quote was from 2002, a year by which scientists should all have noticed that females' heads are likewise huge.

We know now that the high prow of the sperm whale's head is really the living world's greatest sonic boom box. Almost the entire head of the sperm whale is a factory of vibration and amplification. Compared to a dolphin's sonar hardware, the sperm whale's is supersized. Even as a fraction of body weight, it's still about twenty times as large as a dolphin's relatively meager investment in the area known as its "melon."

Unlike our skull, which is rounded at the forehead, toothed whales' fore-skulls are dished into a bony sound reflector. That dished-out foreskull is directly above the whale's eyes, as your own bony forehead is above your eyes. The biggest part of a sperm whale's ocean-splitting wedge of head, extending far forward of the skull, contains no bones *at all*. "The Sperm Whale's head," Melville informs us, ". . . is of a boneless toughness, inestimable by any man who has not handled it. The severest pointed harpoon, the sharpest lance darted by the strongest human arm, impotently rebounds from it. It is as though the forehead of the Sperm Whale were paved with horses' hoofs."

To generate their sonar, they force air across structures called "phonic lips" in the air passage inside the blowhole. A sperm whale's blowhole is, weirdly, their left nostril. That's why they blow a distinctly left leaning spout from the tip of the "snout." The *right* nasal passage does not exit the whale's head at all. Its sole function is pushing air through those phonic lips. This creates vibration. That's the beginning of how a click is generated.

Now the vibration enters the fatty organ that gives sonar-enabled whales and dolphins that high, rounded outer forehead. Aptly called the melon in dolphins, this organ uses lipids of different densities as sound lenses. The energy travels through the oil-filled "spermaceti" organ that is most of the upper part of this whale's head. It reflects off an air sac immediately in front of the great bony dishlike front of the whale's skull. The sound then passes through a series of acoustic lenses in the lower half of the whale's vast bulb of a head, which is a gigantic sound-amplifying system. This series of reflections and focusings of the vibrational energy amplifies and sharpens the click. (Whale hunters, wholly ignorant of any of this and unable to find much profit in these astonishing sound-amplifying lenses, called this anatomical region "the junk.") What emits through the skin at the forefront of the whale's head is a weapon of sound. These are sharp broadband clicks with energy between 5 and 25 kilohertz.

Ahab's cry of "Forehead to forehead I meet thee . . . Moby Dick!" would lose its symmetry and its alliterative ring if written more accurately as "Forehead to nose tip." Author Richard Ellis commented, "If Melville had any idea what really went on inside that nose, *Moby-Dick* would have been a very different novel — wilder and deeper, perhaps, but certainly a lot noisier." Whales would have telegraphed panic, babies would have cried out their pain to their anguished mothers, and struck whales would have turned and buzzed their X-ray-like sonar through whalemen. In fact, all of this did happen. But not in Melville's novel. It happened only in reality whenever the whales were hunted.

The sounds emitted from the whale echo back from potential prey and other objects of inquiry. Just as the sound beam does not emanate from the mouth, the returning echo does not return to "ears." Instead, the echoes are first received by the whale's lower jaw. Whales' specialized, fat-filled lower jawbones set up minute vibrations. It's as if their jaws are also their antennae. The returning vibrations go through something called the "acoustic funnel," which is what it sounds like: a cone-shaped bony structure in direct contact with the acoustic fats inside the jawbone. The funnels channel vibrations to fluid-filled inner ears.

I've been talking about the whale emitting "sound," which is wrong. The whale creates a focused beam of *vibration:* some vibrations bounce off objects, and some of those returning vibrations strike the whale's jaw. Neurons change the vibrations into nerve impulses that enter the brain. Sound is a perception rather than a property. The brain *creates* "sounds" as an *interpretation* of nerve impulses. It's analogous to how a speaker or

115

video monitor creates music and images from digital or analog impulses. (In fact, the brain doesn't even require new impulses to create a subjective sensation; that's why you can "hear" your favorite song in your head, and why you sometimes can't get an unwelcome song out of your head.) The brain also creates other perceived experiences, such as color and visualizations, for that matter.

Eyes and ears collect samples of the waves and vibrations. Connected nerves convert them into impulses. Brains translate the impulses and analyze them. So far this is pretty mechanical. Cameras and sonar machines do all of the above. The miracle occurs when the brain displays its analysis and we *experience* the conscious sensations of seeing the world, or — as I'm doing right now while writing this — hearing Beethoven's Seventh Symphony through my computer and effortlessly experiencing the different sounds of the strings, horns, and drums, along with the psychological effect of their rousing magnificence. No one understands how neuronal processes in our brains result in *felt experience,* our actual sensations.

That vast frontier of a head that houses the sperm whale's sonar factory also houses one other thing: the universe's largest known brain. About twenty pounds. (Fellow humans: ours clocks in at three.) It weighs more than the brains of whales whose bodies are twice

116

as large. Perhaps the sperm whale's brain is big even for a whale because a very social whale needs a big brain to keep track of many individuals. Perhaps it is outsized because the brain is the central processor of all those faint echoes returning from the deep ocean's frigid blackness. All of this brings us to an interesting possibility: Might whales (and bats) — use sonar to actually *see*? Think of it this way: hearing and vision are both created from energy *reflected* off objects. We see *echoes* of light. Why not see echoes of sound? Commercial sonar sets available in any boating-supplies store bounce vibrations off objects, detect the returning echoes, and send those impulses through wires to a display screen, where the returning "sound" is interpreted into colored pictures. The fact that the human brain visualizes echoes of certain electromagnetic waves (so-called visible light) but hears vibration echoes seems arbitrary. What prevents a brain from visualizing both? Some blind people have apparently acquired impressive skills at using echoes from their own clicking sounds to navigate using what's called echolocation; they say the sound lets them form actual images. Scientists are finding that what's been thought of as a "visual" part of the brain is really a spatial part, and it doesn't much matter whether the input comes through eyes or ears. They are coming to believe that "the brain is organized by task rather than sensory modality . . . the primary 'visual' cortex can be adapted to map spatial locations of sound in blind humans" and that "stimulus maps for sound in expert echo-locators are directly comparable to those for vision in sighted people."

Might sonar-using whales, dolphins, and bats process the echoes into visual percep-tion, experienced as images? It's difficult for us to conceive of a dolphin streaking through the sea at night in pursuit of fish, or a bat swerving sharply to catch a moth, or a sperm whale locating squid in total darkness without imagining that they are somehow visualizing their world. What we know for certain is that they are astonishingly good at knowing where and what things are, at very fine resolution and very high speed.

Sperm whales require the greatest sonar device in the living world. Otherwise, evolu-tion would never have requisitioned so much mass, engineered so much structure, and compromised so much efficiency of move-ment for this purpose, or devoted to it so much energy, neural configuration, behavioral instinct, learning ability, and time. Sperm whales spend about 80 percent of their lives propelling that detection device through frigid blackness while ticking every half second or so. Lifetime, well over a billion clicks.

In *Peter Pan,* the crocodile's menacing

water-shrouded approach is detected by the ticking of the clock he swallowed in addition to Captain Hook's hand. In the real world, a much greater being clicks through the darkness in all seriousness. At every moment of our own lives and loves, during all times and tides and all our work and leisure, sperm whales by the thousands, clicking their queries into the unbounded spaces of the deep world, are traveling with their families.

I mentioned that males leave their families at around adolescence. Male and female sperm whales differ mightily in both lifestyle and size. Males can reach nearly double the length and triple the mass of adult females. Females top out at 35 to 40 feet (11 or 12 meters or so) in length and about 15 metric tons (33,000 pounds). Males are much larger, at around 65 feet in length (20 meters) and perhaps 45 metric tons (nearly 100,000 pounds). Even larger males have swum the seas. The whale who rammed the *Essex* was estimated at 80 feet. The largest ever measured appears to be the 84-footer killed off Japan in the early 1800s and recorded by Beale (I've adjusted his punctuation for clarity):

Of the largest size or about eighty-four feet in length, the dimensions may be given as follows: depth of head from eight to nine feet; breadth, from five to six feet; depth of body seldom exceeds twelve or fourteen feet, so that the circumference of the largest sperm whale of eighty or eighty-four feet will seldom exceed thirty-six feet. The swimming paws or fins are about six feet long and three broad. . . . Flukes: they are about six or eight feet in length, and from twelve to fourteen in breadth in the largest males. . . . When we consider these as applied to the circulation, and figure to ourselves that probably ten or fifteen gallons of blood are thrown out at one stroke, and moved with an immense velocity through a tube of a foot in diameter, the whole idea fills the mind with wonder.

Males live with their mothers for about ten years. As with elephants, male sperm whale separation from their birth families happens gradually, the adolescent taking as long as half a dozen years to finally swim into the deep indigo for good.

Shane explains that researchers used to think that adolescent males left their families because puberty's testosterone motivated them to go off on their own. "But what really happens," he says, "is sort of sad." When a mother has a newborn, she no longer wants her adolescent son around. The other adult females don't want to socialize with their maturing nephew or grandson, either. At around ten to fourteen years of age, a young male gets socially isolated out of the family.

Shane's group was the first to follow individually known sperm whales long enough to see that process. "Pinchy had a young male named Scar. But after Pinchy gave birth to Tweak, none of the adults showed any further interest in Scar. Tweak, the baby, was the only family member who wanted to spend any time with Scar. You'd see Tweak playing with Scar, and Pinchy would come and just take Tweak away."

In some cases, the adolescent male will follow his family around for a couple of years before it seems as though he finally decides, "Well, if no one will hang out with me, I guess I'll leave."

It's possible that the final split happens when an adult male comes to visit and a young male leaves with the mature one. Whichever way it goes, all-male groups roam truly vast distances. Occasionally males even swim from one ocean to another. Adolescent males join up in temperate waters, their herds like voyaging frat houses. With age, mature males push farther away from companionship, moving to higher and higher latitudes, colder and colder solitudes. Like elephants, again, they probably don't breed for some ten to fifteen years after leaving their family group, when they're around thirty. To do so, they retreat from their chilly summer haunts and make breeding excursions to warm-water hookups.

So we have females and young swimming in tropical and subtropical oceans, while adult males — especially until they begin breeding — spend time mainly in cold-water regions.

When males come to the tropics to consort, they are magnets to females. Males can make a loud "clang" that apparently rings for romance. It is *much* louder and more resonant than their regular clicks, and they repeat this come-a-courting bell about every five to eight seconds. Researchers believe that sperm whales can hear these clangs from as far away as sixty kilometers (nearly forty miles). Males can draw dozens of females to them. And unlike females observing clan divisions, males interact freely with females of various clans. Roaming males thus keep the gene pool well stirred.

Jonathan Gordon recorded his first impressions of male-female interactions, writing, "I had expected these huge males to be forcing their attentions on unwilling females; what I observed under water could not have been more different. The male was the focus of intense attention from all group members." Mature females, juveniles, and youngsters all rubbed and rolled along the male's body. "They just seemed delighted that he was there," Gordon continued. "For his part the male was all calm serenity and gentleness." So much so that on several occasions a male has been seen gently holding a youngster in his mouth, as though greeting the baby affectionately.

Mating seems to be up to the females. We don't yet know what sperm whales consider sexy. If several males appear, the females may get interested in just one particular male, and not necessarily the biggest. Sometimes a very large male will appear but the females won't seem to care.

After breeding, the sires voyage along, vanishing into the blue-dark infinities. "My Lord Whale has no taste for the nursery, however much for the bower," commented Melville, "and so, being a great traveler, he leaves his anonymous babies all over the world."

The implications regarding sperm whale culture are stark and definitive. That well-mixed DNA proves that sperm whale clan differences, and their sense of identity, are not genetic. They're entirely learned. Clans are both the learned attraction of certain families to one another and the acquired aversion of certain families for one another. In a gene pool sense, sperm whales are basically one genetic "stock." But *culturally* they are their own mosaic of learned traditions. Each clan is one tile in that mosaic.

FAMILIES:
SIX

Far above all other hunted whales, his is an unwritten life.

— Herman Melville, *Moby-Dick*

Sophocles arrives back in our world with a sudden explosive leap and terrific crash, like a school bus bursting from the ocean. Wrenching that huge squared-off head and dark-bodied bulk clear out of the ocean seems impossible — except that they can, and they do, and she just did. This new afternoon promises to be a good one.

A few minutes later, *PHOOOSH!:* Jocasta pops up close to us, relieving herself of the pent-up, reexpanded gases in her spent lungs.

Now Laius is up too, half a kilometer from Jocasta. To them, this is no distance at all, and after about ten minutes they simultaneously fluke up and dive.

An hour later, Jocasta and Laius surface together 3.6 kilometers from where they dived. Add the distance down and the distance up: they've swum about five and a half kilometers.

All afternoon, young Jonah had seemed alone. But Jonah has been tracking her mother, Sophocles, from the surface, and reuniting with her for part of every hour.

And so here's the thing: in the ocean vastness, maintaining family cohesion requires constant effort. It is *intentional.*

For being a sperm whale there's no instruction manual, no rule book. There are only demands and generalities, rhythms and patterns. Mostly there is this: the wide, deep ocean and family bonds.

When you hear the word "whale," you think "heavy." But a whale in water is quite literally weightless. Whales have enormous mass. In the ocean, mass has no direct relation to weight. Land applies gravity. Ocean adds buoyancy. On land, mass correlates with weight, but in water, mass correlates with distance. In water, a whale can fly. Compared to a smaller being, a large ocean animal covers longer distances with each propellant thrust. That translates into ease of traveling from surface to depths, huge home ranges, and horizon-hungry migrations.

Sperm whales live *rhythmic* existences, clenched within dynamic tensions of opposing forces. On an hourly basis, they live *vertical* lives, yo-yoing as they do between a fresh, bright breath of air and frigid, lung-crushing darkness half a mile down. Light and warmth, air and breath are up. Babies are up. Food is down. The greatest daily rhythm maker in our human lives — the day-night cycle — doesn't matter much to these whales. During the bright sunshine of our days, the whales spend most of their hours searching inside perpetual darkness. But they do tend to socialize around sunset and during nighttime. Researchers sometimes call such occasions "teatime." That's when the whales come together and make codas and play, reinforcing social bonds by touching, rolling around.

"Those are moments you really value," Shane offers. "You feel very lucky to be there with them."

I saw it once, in the Gulf of California. That group had a newborn so minutes-fresh that its tail hadn't fully unfurled and its umbilicus remained attached. Everyone in the group lingered at the surface, touchingly close to one another. They were like that when we first spotted them and remained like that when we left them in peace.

Such seemingly joyful group-reinforcing events remind me of the greetings I've seen during the energetic bonding ceremonies of elephant family members, when they're trumpeting and twining their trunks with family and friends. Or African lions waking from a long rest and rubbing up against each other like the huge cats they are. Or the kinetic face-licking, tail-waving rally of a wolf

126

pack. Or just the way our dogs greet us in the morning.

They don't simply show that they "identify" each other. They demonstrate that — this is the only way to describe it — they are excited to be together. Are they happy? Feeling good? Glad to be with their familiars, to whom they are emotionally bonded? No other explanation fits the evidence or explains the performance. Nothing else I can think of begins to account for the existence and function of such emotions, except that they occur for a reason.

At two-thirty p.m., two and a half hours after leaving Jocasta's family, we've traveled southwest about nine kilometers, or five and a half miles. Shane is hearing more sperm whales. He thinks they're a different family.

Most days when there are whales here, off the west shore of Dominica, it's only one family unit at a time. Shane *has* seen as many as six families here at once, about thirty whales. That time, a male had arrived, inciting that convention. It must have been a good day for all.

But whomever Shane's hearing at the moment, they're faint. And deep. So we move toward where the sound seems a little stronger.

This whale-sized cat-and-mouse appeals to me. It's interesting. It's a science technique.

It's a hunt. It's a game. It is metaphor. We all yearn to make the right next choice, to put ourselves in a good place, or near enough to good, for now and for what's coming.

On the next stop, the clicks are clear. Whales are near. Suddenly they stop. From under his hat and sunglasses and earphones, Shane says, "Corners everyone."

And so we station ourselves around the boat, eyes outward.

And in under a minute, *"Blow!"* Just fifty yards off the bow.

When this whale flukes up, Shane instantly exclaims, "Mrs. Right!" The R Family also includes Rip and her young one Rap, Raucous and her young one Riot, Rita, Rema, and Roger. Named after the affirmation "Roger that," Roger is female.

Two other whales we'd been hearing come up half a mile apart and begin puffing their white clouds of breath as though smoking peace pipes. They are Sally of the S's and Roger of the R Family. Two families known to like each other. Members of the same clan, of course.

That sperm whales will socialize only with other families who share their coda dialect implies that those codas are *markers* of deeper differences between clans. I wonder aloud about what these might be.

"The *underlying* differences?" Shane asks, repeating my query. "That's an untouchable

128

question right now."

Camaraderie and group identity can have negative poles of fear and segregation. Somewhere below that visceral response there is an answer to the question "Why?" But often — too often — even when it's about our own species, we don't know.

"It matters most when you don't know anyone," Shane says. "You must decide: 'Am I going to socialize with you, cooperate with you, avoid you — ?' "

The things that make group differences *matter* involve knowing what's expected, understanding what the group does, and, ultimately, being able to join in. When "birds of a feather flock together," there's a reason. The question isn't entirely untouchable. As Shane explains it, it seems rooted in questions of cooperation. Cooperation depends on shared expectations. Expectations allow us to cooperate. But between cultures, expectations differ. For instance, if I work for you, I expect you to pay me. If I expect money, and you expect to simply return the favor at a later date, there will be trouble. If we both expect *either* an exchange of money *or* a bartering of favors, no problem. Members of a group must share similar expectations.

Cultural expectations may be entirely arbitrary. When greeting, you're expected to either shake hands or bow; when eating, to use a fork or fingers or chopsticks; when meeting, to offer your business card a certain way; in the evening, to wear something within the bounds of your culture. Often the only reason is "It's just how we've always done things around here." Arbitrary little differences can make it impossible to function in a group that's not your own. Knowing what's expected keeps group members together and keeps different groups separate and apart.

Among sperm whales, differences in groups' behavior translate to differences in survival. Off the Galápagos Islands, one clan that was studied — the Regulars — stuck closer to shore and meandered, moving about a dozen miles daily, around twenty kilometers. The Plus-Ones, by contrast, breezed past the islands farther offshore, in rather straight shots that displaced them by more than thirty kilometers. The straighter-moving clan has better foraging success in "El Niño" years, when food is sparse. But in other years, the meandering clan finds its advantage. (El Niño is an old name for a shift in prevailing tropical Pacific winds and currents. It greatly changes the ocean's productivity and affects weather around the world.)

Here off Dominica, too, one of the two clans tends to swim closer to shore. To the question "How do we live in this place?," each group passes to its young its particular answers about what to say, where to live, how to move, how to hunt. When killer whales in one population learn how to catch stingrays and those in another population teach their children how to lunge through the surf to snatch sea lions off a beach, it's a bit analogous to how human children in many cultures pick up their elders' skills at farming, fishing, the family business, organized crime, or what have you. In a word, they've learned the *specialized* ways of their elders. Cultural specialization, by keeping some together, some apart, builds diversity.

And that touches a major implication. As culture influences who survives and who doesn't, genes that facilitate cultural learning get favored. Before culture, genes dictated all the behaviors. After culture, behaviors begin to dictate which genes are working. As genes affect learning, learning affects survival, and the survival advantages of learned culture result in the spread of genes that facilitate cultural learning. The species becomes more culturally inclined.

So crossing the culture horizon is a really big threshold. It's got big evolutionary consequences. The world changes living things. Sometimes it changes them in ways that let living things change the world. Differences as superficial as different vocal traditions — analogous to different human languages — can sometimes, over time, lead to change that is deep and permanent.

Let's say one group learns a specialty and

another group learns a different specialty. Their techniques for foraging, say, are very different. Say the specialist groups avoid each other. Cultural segregation prompts further specialization. Then the specializations lead to genetic evolution and actual physical changes. In that scenario, culture leads, and genes follow. Eventually, evolution might drive specialist populations into separate species.

This is exactly what's happening in killer whales. The fish-eating orcas and the mammal-hunting orcas of the Pacific Northwest transit the same coasts and coves — but scrupulously avoid each other. They have not interbred for tens of thousands of years. During those generations, the mammal hunters have been evolving more robust jaws and teeth to subdue their considerably more dangerous prey. Except in name (scientists named them all *Orcinus orca* — before they understood all this), they are in fact separate species.

Globally, the world's ten or so "types" of killer whales differ mainly by their food habits. Their pickiness is almost incredible. One wants one species of salmon. Another specializes in hunting sharks. In the Antarctic, one type mainly hunts minke whales. Another dines on penguins. And Antarctic seal eaters want Weddell seals, passing up the crabeater seals that make up almost 85 percent of the seal population there. They're so finicky that when they spot a seal on ice, they identify the species before they bother to call their friends. If it's one or more Weddell seals, the whales will perform a specialized technique they have learned of swimming in unison and creating a tail wave that washes over the ice, spilling the seals into the sea. Other killer whales catch tiny, agile herring by expertly herding the fish into a tight ball that huddles against the surface, then tail-slapping them and slurping up the stunned fish amid a glitter of drifting scales. These hidden cultural differences have sent killer whale cultural groups on differing evolutionary trajectories worldwide. We tend to think of culture as almost the *opposite* of genetic evolution. But in many species, culture-based segregation has fundamental evolutionary implications that have been almost entirely overlooked by scientists.

Dolphin communities also often contain hidden structure. Many dolphin groups specialize in one technique or another, clearly learning from and sharing with their children and friends. Some use one specialized individual "driver" to herd fish toward the others, or to raise a ring-shaped cloud of mud around schools of small fishes. In the Adriatic Sea, two groups of dolphins "time-share" the same area, using it on different schedules. One of those groups follows fishing trawlers for their discarded catches; the other never does. Off Brisbane, Australia, the arrival of shrimp trawlers caused a cultural split: 154 bottlenose dolphins began following for discarded fish; 88 other dolphins in the same area did not. The two groups began avoiding each other, creating two new social classes. When the boats depleted their prey and abandoned the area, the panhandling dolphins went back to hunting *and* they all went back to mixing and socializing.

Tendencies of communities to "stick with what you know" seem to create "resident" and "non-resident" groups. In South Carolina, the pioneering conservation biologists Sally and Tom Murphy took me to watch resident bottlenose dolphins working in small groups to strand small fishes on muddy banks. Only residents use this traditional technique; dolphins who occasionally pass through the area never do. They're not just dolphins. They're particular dolphin groups who do particular things.

Sticking to one's group can offer efficiency advantages simply because everyone knows what to expect and how to cooperate. If you're a killer whale hunting fish, you want to be in a big noisy group that will scare fish into schooling up tightly, but a group hunting sea lions must deploy in small, stealthy, silent packs. The two strategies are not compatible, so the specialists must not mix.

So they don't.

Nowadays in our globalized, high-density, multiracial, polyglot society, there's a lot at stake in acceptance of differences, diversity, other traditions and cultures. But our evolved brain resists, sometimes violently. Our evolved brain is best adapted to social life in a vast wilderness where human groups are small and familiar, and strangers can mean instant trouble. Our evolved brain advises us that because we must cooperate with others to survive, it pays to avoid those with different languages or customs. People want what's familiar.

The basics of being human are similar across cultures. But the details vary so importantly that the differences could be lethal. Imagine a woman raised in Los Angeles who works in a bank being suddenly transplanted to a traditional Inuit village where the language is impenetrable to her and surviving depends on managing dogs, making leather, and hunting seals. Imagine an Inuit in his parka dropped behind the teller window of a California bank. Imagine a desert San ("Bushmen") camp where drinking water must be squeezed from particular plants. If you grew up hunting monkeys with a blowpipe, you'd be a liability among harpoon-wielding walrus hunters. People thrust into a sufficiently different culture would have so rough a time, they'd quite possibly die. And that's if they aren't shunned. Yet for those born and raised in the local knowledge — whether it's how to acquire a credit card or how to haft a spearhead — the culture works. The culture is home.

Self-imposed segregation begins because it's easier to be with those you understand. That's how it often is for people. And for sperm whales. It's hard to imagine how, if sticking with your group didn't have survival implications, a system of clans could have developed. But it did. Knowing who you are, and who is who. Knowing *the way* "we" do things. Thus equipped, sperm whales take their clan membership voyaging through the worldwide deeps.

FAMILIES:
SEVEN

When by chance these precious parts in a nursing whale are cut by the hunter's lance, the mother's pouring milk and blood . . . discolor the sea for rods.

— Herman Melville, *Moby-Dick*

When I tell Shane that someone I know claims to have seen a white sperm whale while sailing through the central North Atlantic, he replies matter-of-factly, saying, Yes, that whale is well known to researchers around the Azores. White humpback whales, too, have been photographed and named. And during two centuries of commercial whale hunting, several white sperm whales were encountered, with the consequence you'd expect.

Herman Melville read about one, named Mocha Dick for his frequented haunt off Mocha Island, Chile. That "white as wool" whale had reportedly escaped dozens of violent encounters with whalers. One J. N. Reynolds penned and published the story of

137

the killing of Mocha Dick as told to him by the first mate of a whaling vessel in the vicinity of Mocha Island. The man, in his mid-thirties (Reynolds does not supply his name), may have exaggerated certain details. But the existence of the whale appears real enough.

The first mate tells Reynolds that several harpoon boats had been lowered from the ship and the crews had been rowing hard after a group of sperm whales for about a mile when the whales all went down. The men stopped rowing. After about five minutes, they saw what proved to be a young, nursing-age sperm whale "playing in the sunshine." At that age, the whale was too young to follow the others when they went deep. The whale hunters knew enough about mother culture to exploit it. The person speaking is the first mate himself:

> "Pull up and strike it," said I to the third mate; "it may bring up the old one — perhaps the whole school."
> And so it did, with a vengeance! . . . Hardly had it made its first agonized plunge, when an enormous cow-whale rose close beside her wounded offspring. Her first endeavor was to take it under her fin, in order to bear it away; and nothing could be more striking than the maternal tenderness she manifested in her exertions to accomplish this object. But the poor thing was

dying, and while she vainly tried to induce it to accompany her, it rolled over, and floated dead at her side. Perceiving it to be beyond the reach of her caresses, she turned to wreak her vengeance on its slayers, and made directly for the boat, crashing her vast jaws the while, in a paroxysm of rage. Ordering his boat-steerer aft, the mate sprang forward, cut the line loose from the calf, and then snatched from the crotch the remaining iron, which he plunged with his gathered strength into the body of the mother, as the boat sheered off to avoid her. . . . At that instant, a whale "breached" at the distance of about a mile from us . . . my old acquaintance, Mocha Dick . . . white as a snow-drift!

. . . No sooner was his vast square head lifted from the sea, than he charged down upon us, scattering the billows into spray as he advanced, and leaving a wake of foam . . . as he cleft his way toward the very spot where the calf had been killed. "Here, harpooner, steer the boat and let me dart!" I exclaimed, as I leaped into the bows . . . though he were Beelzebub himself! . . . I raised the harpoon above my head . . . and sent it, hissing, deep into his thick white side! . . . The instant the steel quivered in his body, the wounded leviathan plunged his head beneath the surface, and whirling around with great velocity, smote the sea

139

violently, with fin and fluke, in a convulsion of rage and pain.

Mocha Dick dives at such speed that hundreds of fathoms of line buzz through the boat's smoking chocks. But "overpowered by his wounds, and exhausted by his exertions . . . the immense creature was compelled to turn once more upward for a fresh supply of air." He blasts through the surface, tows the boat along for another quarter mile, then abruptly stops, "and lay as if paralyzed . . . quivering and twitching." The whale lunges for the boat, but collapses.

Because the men succeeded in killing the famed Mocha Dick, that tale is recounted in the victor's triumphal timbre. More soberly documented — because of the appalling consequences for the ship and men — was the enraged whale who sank the Nantucket-based whaling ship *Essex* in 1820.

During Melville's voyage on the whaler *Acushnet,* he chanced to meet William Chase, who was among the crew of another whale-hunting ship. William was the son of Owen Chase, who'd witnessed the whale's attacks on the *Essex* and had survived the horrifying misery in the lifeboats during the three months that followed. William lent Melville his father's account, titled *Narrative of the Most Extraordinary and Distressing Shipwreck of the Whale-Ship Essex.* Melville later jotted in his own copy, "The reading of this wondrous story upon the landless sea, and so close to the very latitude of the shipwreck had a surprising effect on me."

The *Essex*'s crew had found their planned hunting area off western South America already depleted. Consequently, the twenty-nine-year-old captain, George Pollard, followed rumors of whales some twenty-five hundred miles west of where they'd originally intended to search, a vast trek across unfamiliar reaches of ocean. Finally they found whales, and Owen Chase — then twenty-three years old — harpooned a female, whose tail then smacked a leak into their boat. Stanching the leak with clothing, they returned to the ship. Back aboard the *Essex,* Chase saw a huge male at the surface. He blew several times and vanished. When he reappeared, the enraged whale — "as well as I could judge about eighty-five feet in length" — was coming headlong toward the eighty-eight-foot *Essex.* He smashed into the bow and "the ship brought up as suddenly and violently as if she had struck a rock and trembled for a few seconds like a leaf." Smashing the hull's heavy planking must have stunned the whale, for he appeared "apparently in convulsions, on the top of the water about one hundred rods to leeward [roughly five hundred yards] . . . and I could distinctly see him smite his jaws together, as if distracted with rage and fury." But like a stunned boxer on the ropes, the whale seemed to shake the blow off and suddenly remember the fight he was in. The ship was already beginning to sink when, Chase wrote,

I turned around and saw him . . . coming down apparently with twice his ordinary speed and, it appeared to me at the moment, tenfold fury and vengeance. . . . The surf flew in all directions about him, and his course towards us was marked by white foam. . . . His head was about half out of the water, and in that way he came upon us again and struck the ship. . . . His aspect was most horrible and such as indicated resentment. . . . He came directly from the shoal which we had just before entered — and in which we had struck three of his companions — as if he were fired with revenge for their sufferings.

The first strike caused damage sufficient to doom the *Essex*. The second smashed in, or "stove," the front of the ship below the waterline. No one had ever heard of a whale sinking a ship.

Of the *Essex*'s small boats that had been engaged with the whales, Captain Pollard's was the first to return to the sinking ship. Chase wrote: "He stopped about a boat's length off, but had no power to utter a single syllable; he was so completely overpowered with the spectacle before him. He was in a short time, however, enabled to address the inquiry to me, 'My God, Mr. Chase, what is the matter?' I answered, 'We have been stove by a whale.' "

It must have been sickening beyond imagining to watch their home settle deeper and deeper into the vast salt-seared desert that stretched like a drumhead far beyond the rim of all horizons. The crewmen gathered a few things and some bread and water and cast themselves and their fates into three little whaleboats. They headed *not* to the nearest land — Tahiti — because, in ignorance, they erroneously believed they'd fall prey to cannibals. Ironic in the extreme, their dread not only sealed death for most, but ensured that the survivors would become what they'd worst feared.

Three months later, having wandered over some forty-five hundred miles, of the twenty who left the wreck of the *Essex,* eight barely alive men were plucked up by a British frigate. Their gruesome sufferings and the cannibalism that kept those eight alive are recounted in Nathaniel Philbrick's *In the Heart of the Sea,* based in part on Owen Chase's narrative and on the diary of a surviving fourteen-year-old cabin boy named Thomas Nickerson.

The *Essex* sank in 1820, when Herman Melville was one year old. Chase wrote his account in 1821. Mocha Dick was killed in 1810, but the account was not published until 1839. In that year, Thomas Beale brought forth his excellent *Natural History of the Sperm Whale.* In 1841, after Melville met Chase's son William, Melville deserted the *Acushnet* in the Marquesas and worked his way home; *Moby-Dick* was published in 1851.

Also in 1851, off the Galápagos Islands, a sperm whale bit and smashed to splinters the whaleboat from which he'd just been harpooned, before ramming a puncture into the ship *Ann Alexander,* sinking her. And in 1902, a thousand miles east of Brazil in the South Atlantic, while the *Kathleen*'s three boats were all busy dealing with whales they'd harpooned, a free whale struck the ship, also inflicting damage sufficient to sink it.

Richard Ellis says of all this, "We are at a loss to assign a motive for a whale's attack. . . . We simply don't have an idea of how the sperm whale thinks — or, for that matter, whether this animal, with its gigantic brain, thinks at all."

That sentiment reflects a common and profound incapacity in our relationship with other animals. All the whales who ever attacked whaleboats or sank ships had just been harpooned or had just witnessed their companions under attack. Is it really difficult to

144

see a pattern? Their motive: defense. Self-defense, defense of their group.

They'd been engaged in the business of being themselves when they came under assault by humans from half a world away, who might as well have been alien beings in a spaceship. Sperm whales are peaceful creatures, but they live in an ocean where dangers lurk. They understand aggression, and so they understand defense.

The most salient point regarding whales and their culture is that both Mocha Dick and the whale who sank the *Essex,* as well as, possibly, *Kathleen*'s scuttler, were huge male sperm whales who freely came to the defense of females and young under attack. Because male sperm whales form no pair bond and thus appear to have little long-term stake in any particular group, these are perhaps the purest examples, in species other than humans, of altruism.

Sperm whales don't bother each other, other sea creatures (except their food), or humans. People nowadays have learned that they can swim with them with no worry other than whether they'll get good photos.

For centuries, most meetings between sperm whales and humans did not end well. Here is Beale's description of a sperm whale in contact with men:

145

Mad with the agony which he endures. . . . Suffering from suffocation, or some other stoppage of some important organ, the whole strength of its enormous frame is set in motion for a few seconds, when his convulsions throw him into a hundred different contortions of the most violent description, by which the sea is beaten into foam. . . . And the mighty rencontre is finished by the gigantic animal rolling over on its side, and floating an inanimate mass on the surface of the crystal deep, — a victim to the tyranny and selfishness, as well as a wonderful proof of the great power of the *mind* of man.

Why a harpooned whale or a companion might occasionally mount an aggressive defense is not too hard to understand. One might less successfully attempt to fathom "the great power of the *mind* of man."

FAMILIES:
EIGHT

All sperm whales, both large and small, have some method of communicating by signals to each other, by which they become apprised of the approach of danger, and this they do, although the distance may be very considerable between them, sometimes amounting to four, five, or even seven miles. The mode by which this is effected, remains a curious secret.

— Thomas Beale, 1839

More than a century and a half after Beale, the communication abilities of sperm whales under attack continued to astonish scientists. But sperm whales themselves understand that in their culture they can summon help, and that it can be important to go to the aid of the distressed. Whale experts Bob Pitman, Lisa Ballance, Sarah Mesnick, and Susan Chivers have described sperm whales responding to killer whales during several interactions.

One day they were watching two groups of

sperm whales when they spotted five killer whales heading toward the second group. The sperm whales of the second group then submerged, leaving a baby at the surface for less than a minute before they resurfaced. The researchers noted,

> We think they may have sounded an alarm call at this time because immediately afterward the first sperm whale group bunched up and started traveling rapidly toward the second group. When the second group resurfaced it was joined almost immediately by several other sperm whales that surfaced around and among them, and by the time the two original groups merged there were approximately 15 individuals present.

Some of the sperm whales poked their heads out, looking around, and some slammed the surface with their flukes. A single adult female killer whale advanced among the sperm whales, and an oily slick that formed at the edge of the sperm whale cluster suggested that she had bitten one of them. But more and more sperm whales continued to arrive:

> As their numbers increased to approximately 20 . . . at least four other groups of sperm whales in the distance were charging toward the core group at full speed, push-

ing waves with their heads as they plowed through the water. The converging animals initially included separate groups of eight, five, and two animals, and a large, lone individual that appeared to be an adult male.

As the scientists continued observing whales hurrying toward danger, the sperm whale group, augmented with the new arrivals, created an almost military parade formation, "from one to four or five animals wide and 12 to 15 animals long, and all facing the same direction." In this formation, they came about-face several times, usually all turning in the same direction, always clockwise.

When the female killer whale moved back in among the sperm whales, "causing much agitation," nearly all the sperm whales submerged. Less than a minute later, when they resurfaced, approximately thirty were present. They'd just picked up *ten more whales.*

Pitman and his colleagues noted that "every sperm whale within at least a 7-km radius immediately charged toward the threatened group at high speed and joined them in a defensive formation. . . . There can be little doubt but that they were responding to a very specific and powerful acoustic signal." (The scientists could see all this with giant swivel-mounted "big-eye" binoculars, situated on the bridge deck high above the waterline, affording them the ability to sight whales over quite a few miles to the extended horizon.) Then the sperm whales "formed a staggered chorus-line formation with the entire group lined up, facing the same direction, side-by-side, apparently touching each other," a "remarkably precise formation."

Almost incredibly, *more* sperm whales continued to appear and join, until the group swelled to an estimated fifty. As Shane had said, "When killer whales show up, it seems like sperm whales suddenly come popping up out of nowhere."

The female killer whale was last seen traveling almost two kilometers away, toward the rest of her group, none of whom had at any point decided to attempt an attack. That's the way sperm whale mutual defense is supposed to work.

In the northern Gulf of Mexico in 2011, researchers saw five killer whales harassing a group of nineteen sperm whales with two young ones. They did not attack. In another event in the gulf, sperm whales fending off harassing pilot whales made their defensive formation a three-dimensional rosette, with whales' heads near the surface and their tails downward — and young whales in the protected center. Sometimes two adults got beneath a young one, keeping the juvenile between them and the sea surface above. The pilot whales left. These events show that the ocean is a dangerous place for sperm whales caught without their group, and that safety requires sperm whales to create a coordinated response. And they do. Usually it works.

But not always, and the difference might be whether there are experienced whales who can lead the group's defense. The same scientists (Robert Pitman and his colleagues) who watched the sperm whales gathering from miles around to repel killer whales witnessed a very different incident at another time and place in which more than two dozen killer whales attacked a family group of nine sperm whales who appeared helpless. The nine seemingly helpless sperm whales went into a rosette formation, but to no avail. Several times, killer whales dragged one of the sperm whales out of their defensive position. Each time, one or two sperm whales left the formation almost immediately, and, despite the catastrophically injurious attacks they brought upon themselves this way, they flanked the isolated individual and led her back into the formation. The killer whales quit attacking once they had killed one of the sperm whales.

The researchers wrote tellingly, "If the sperm whales had not continued to lead separated individuals back into the rosette and instead let the killer whales have one of their number, the rest of the herd might have been spared. Although altruistic behavior likely serves sperm whales well in most cases, in this instance it meant many, if not all, of the herd was sacrificed in efforts to protect individual members." The researchers thought it "quite possible that the entire herd [later] died" from the devastating injuries they'd all received.

In the other incidents I described, sperm whales were quite capable of effective defense, making them long considered rather immune to killer whale attacks. In this episode, though, Pitman and his colleagues were struck by "the apparent helplessness" of the sperm whales. They recruited no aid, did not strike at the killer whales, and failed to attempt to escape by diving deep. What was different in this instance? The scientists saw none of the callused patches that adult females usually bear, "so it is possible," they ventured, that this was "a group of large subadults." Perhaps then, the crucial life-or-death difference was the absence of a wise elder to demonstrate how an effective response to an attack gets done in sperm whale culture.

When sperm whales head toward danger, do they feel that it's the right thing to do? What's clear is that the sperm whales know not just that there is safety in numbers but that numbers mean safety for all.

The question therefore becomes, Can they understand this well enough to form permanent large groups when the threats grow constant? It seems so, as we're about to see. Over the two centuries during which human attacks began, intensified, and then de-escalated, sperm whales first lived in small groups, then traveled in large groups, and now again live in smaller groups. In recent years, in all oceans, sperm whale groups vary from three whales to about two dozen individuals, averaging about a dozen.

But in Reynolds's account about Mocha Dick, in 1810 the whalers came upon so large a group that they first sighted "spouts from a hundred spiracles." Herman Melville, perhaps the most empathic firsthand chronicler of whaling under sail, believed that beleaguered remnants of besieged pods of whales boosted their defenses by joining into large, permanent companies. Melville readies us for a scene of high intensity with this:

> Owing to the unwearied activity with which of late they have been hunted over all four oceans, the Sperm Whales, instead of almost invariably sailing in small detached companies, as in former times, are now frequently met with in extensive herds, sometimes embracing so great a multitude, that it would almost seem as if numerous nations of them had sworn solemn league and covenant for mutual assistance and protection. . . . Even in the best cruising

grounds, you may now sometimes sail for weeks and months together, without being greeted by a single spout; and then be suddenly saluted by what sometimes seems thousands on thousands.

With that preparation, we enter the scene as several whaleboats have begun to attack the outer edge of a large group of whales and those whales at the edges have stopped to create numerous defensive formations. Meanwhile, in the eye of the storm Melville shows, with great pathos, the whales' innocence:

We glided between two whales into the innermost heart of the shoal, as if . . . we had slid into a serene valley lake. . . . In the distracted distance we beheld the tumults of the outer concentric circles, and saw successive pods of whales, eight or ten in each, swiftly going round and round, like multiplied spans of horses in a ring; and so closely shoulder to shoulder, that a Titanic circus-rider might easily have over-arched the middle ones, and so have gone round on their backs. . . . Keeping at the centre of the lake, we were occasionally visited by small tame cows and calves; the women and children of this routed host.

. . . Possibly, being so young, unsophisticated, and every way innocent and inexperienced; however it may have been, these smaller whales — now and then visiting our becalmed boat from the margin of the lake — evinced a wondrous fearlessness and confidence . . . which it was impossible not to marvel at. Like household dogs they came snuffling round us, right up to our gunwales, and touching them; till it almost seemed that some spell had suddenly domesticated them. Queequeg patted their foreheads; Starbuck scratched their backs with his lance; but fearful of the consequences, for the time refrained from darting it.

But far beneath this wondrous world . . . another and still stranger world met our eyes as we gazed over the side. For, suspended in those watery vaults, floated the forms of the nursing mothers. . . . And as human infants while suckling will calmly and fixedly gaze away from the breast . . . even so did the young of these whales seem looking up towards us, but not at us, as if we were but a bit of Gulf-weed in their new-born sight. Floating on their sides, the mothers also seemed quietly eyeing us. . . .

Meanwhile, as we thus lay entranced, the occasional sudden frantic spectacles in the distance evinced the activity of the other boats . . . carrying on the war.

Parts of that gorgeously rendered scene reverberate with Beale's earlier *Natural His-*

tory; Beale had written, "They lie in the uterus in the form of a bow. . . . The milk, which was tasted by Messrs. Jenner and Ludlow, surgeons at Sudbury, was rich like cow's milk to which cream had been added." Melville turned that into "The unborn whale lies bent like a Tartar's bow" and "The milk is very sweet and rich; it has been tasted by man; it might do well with strawberries."

Melville later informs us that half a century earlier, in the late 1700s, when sperm whales lived in smaller groups (more like they do today), more frequent encounters meant shorter, more profitable voyages. Beale tells us of seeing unimaginably large groups of sperm whales in the 1830s. "I have seen in one school," he relays, "as many as five or six hundred. With . . . always from one to three large 'bulls.' " Melville believed firmly that the whales were responding to the intensified pressure by reconfiguring their social structure, ramping up their social responses to mortal danger.

In former years (the latter part of the last century, say) these Leviathans, in small pods, were encountered much oftener than at present. . . . Because, as has been elsewhere noticed, those whales, influenced by some views to safety, now swim the seas in immense caravans . . . aggregated into vast but widely separated, unfrequent armies.

This amounts to an extraordinary restructuring of social organization, unlike anything else I know of. More impressive, it was accomplished at oceanic scale. How the whales might have decided upon and organized so profound a cultural change, and agreed to maintain it, and later relaxed it remains a complete mystery. But perhaps Melville's observation that it was "as if numerous nations of them had sworn solemn league and covenant for mutual assistance and protection" suggests a formidable function of sperm whales' unique clan culture.

Sperm whales seem to have been essentially unknown to humans before the early 1700s. Living, as they prefer to, in deep open-ocean waters off the shelves of continents, they would not often have been seen on the shallow continental fishing banks that so swarmed with cod. Going after sperm whales seems to have started in New England in 1720. Beale, who was British, tells us that sperm whales were first chased by "a few individuals in America." Few soon became many, and in not many years, "they had not only destroyed great numbers of these useful animals, but had driven the remainder to find more secure retreats, in which they could follow their natural inclinations, without being harassed by the chase or wounded by the harpoon." By 1770 Americans were chasing sperm whales "with extraordinary ardour" in both the North and the South Atlantic. By then, Massachusetts alone was sending 183 vessels to scour the horizon. The steamy geysers of breath they searched for might, in hindsight, be imagined as white flags suing for mercy.

In 1788 a London merchant fitted out, "at a vast expense," a ship to test the speculation that a voyage around Cape Horn into the Pacific might encounter sperm whales in remunerative quantity. *Amelia* sailed from England on the first of September 1788. One year and seven months later, Beale exults, the ship arrived back in England, "bringing home the enormous cargo of 139 tons of sperm oil!"

It took but a few years for whalers to deplete the eastern Pacific. And so, true to script, the boats went farther, longer. *Syren* sailed from England on the third of August 1819 and arrived off the coast of Japan on the fifth of April 1820. There, Beale notes, "she fell in with immense numbers of the spermaceti whale." The crew returned home after an absence of about two years and eight months. During that time, "they had by their industry, courage, and perseverance gathered from the confines of the North Pacific Ocean no less than the enormous quantity of *three hundred and forty-six tons* of sperm oil."

So unprecedented was such success, it "astonished and stimulated to exertion all those engaged in the trade throughout Europe and America."

In 1804, while the Industrial Revolution was still in diapers, one writer intuited where that toddler would take us, and where we would take the whales. In *Histoire Naturelle des Cétacées* (Natural History of Whales), the stunningly prescient Comte Bernard-Germain de Lacépède wrote:

Man, attracted by the treasure that the victory over the whales might afford him, has troubled the peace of their immense solitary abodes, violated their refuges. . . . The war he has made on them has been especially cruel because he has seen that it is large catches that make his commerce prosperous, his industry vital, his sailors numerous, his navigations daring, his pilots experienced, his navies strong and his power great.

Thus it is that these giants among giants have fallen beneath his arms; and because his genius is immortal and his science now imperishable, because he has been able to multiply without limit the imaginings of his mind, they will not cease to be the victims of his interest until they have ceased to exist. In vain do they flee before him; his art will transport him to the ends of the earth; they will find no sanctuary except in nothingness.

Lacépède heard the first rumble, intuited the flash flood, the overwhelming tsunami of us. He was among the first to realize that we would press our withering enterprise to all the far horizons, among the first to perceive, perhaps, and so early in the industrialization of the world, that we would not be inclined to leave room for the rest of life on Earth. By the time Melville sailed, in the 1840s, finding whales even in the western Pacific was already getting more difficult. In consideration, he penned one of the most resonant and haunting inquiries in American literature — one more prescient than he realized:

Owing to the almost omniscient look-outs at the mast-heads of the whale-ships, now penetrating even through Behring's straits, and into the remotest secret drawers and lockers of the world; and the thousand harpoons and lances darted along all continental coasts; the moot point is whether Leviathan can long endure so wide a chase, and so remorseless a havoc; whether he must not at last be exterminated from the waters, and the last whale, like the last man, smoke his last pipe, and then himself evaporate in the final puff.

. . . An irresistible argument would seem furnished, to show that the hunted whale cannot now escape speedy extinction.

But Melville believed that Leviathan *would* likely endure. He seemed to half-hope Providence might more definitively deny humankind the same opportunity, and that Leviathan, who preceded us, would outlast us:

We account the whale immortal in his species, however perishable in his individuality. He swam the seas before the continents broke water. . . . In Noah's flood he despised Noah's Ark; and if ever the world is to be again flooded . . . to kill off its rats, then the eternal whale will still survive, and rearing upon the topmost crest of the equatorial flood, spout his frothed defiance to the skies.

FAMILIES:
NINE

Sperm Whales are not every day encountered; while you may, then, you must kill all you can.

— Herman Melville, *Moby-Dick*

Herman Melville thought that Leviathan *would* outlast "so remorseless a havoc" as the human campaign. But Melville could not have guessed at the intensities that would come. Ahab had vowed to hunt Moby Dick to every corner of the Earth, and in the twentieth century, industrialized nations became a kind of collective Ahab on a maniacal spree.

After petroleum became civilization's major fuel and lubricant, nobody ever again needed to kill whales for light or oil. (Edwin Drake drilled the world's first oil well in Titusville, Pennsylvania, in 1859.) But as whale products were needed less and less, whales were killed more and more. By the 1940s, men were killing cachalots to add their parts to modern inventions such as nitroglycerin and margarine and to lubricate new machines, needs that could have been satisfied with lesser torments.

What had started as rowing men throwing harpoons in a quest for lamp oil and candles morphed into floating factories. Petroleum-fueled ships hunting with cannon-fired explosives had the speed and killing ability to pursue the swifter whales that men pulling on oars could never catch. Blue whales had been inaccessible under sail. Engine-propelled ships killed more than 90 percent of them. Mechanization of every step in the process meant that men could reduce a swimming whale to oil and fertilizer in an hour.

Ten regional whale populations, of five species, were either totally extirpated or so demolished that little or no recovery is evident today, including right whales in the North Atlantic, bowheads in the eastern Arctic, blue whales in the sub-Antarctic, and others.

From the 1700s through the 1800s, sail-driven whalers had killed about three hundred thousand sperm whales. Diesel engines and exploding harpoons allowed twentieth-century whalers to equal the previous two centuries' killing in sixty years, then double *that* in one decade: the 1960s. For all whales everywhere, the twentieth century tallied approximately three million killed.

Into the space age, men intent on killing

whales came up with ever-new uses to justify the unnecessary. Dog and cat food. Fertilizer. Mink food at fur farms. In the 1800s, men had killed sixty million American bison and billions of passenger pigeons, but in terms of pure gross tonnage of animal destruction, twentieth-century whaling beat them all.

The International Whaling Commission was created in 1946 by whale-hunting nations to preserve commercial whale hunting from the impending exterminations caused by: commercial whale hunting. This baked-in conflict doomed it to paralysis. Banal bureaucrats sent to the meetings from whale-hunting nations plodded through charades to "assess" how many whales they would let themselves kill. They were all competing for shares of an overall quota, so a big overall quota meant bigger shares.

Most tortured of the commission's self-delusions was its members' concoction of the "blue whale unit." Countries voted on how many "blue whale units" they would divvy up. One blue whale unit, they decided, would mean either: one blue whale, two fin whales, two and a half humpback whales, or six sei whales. These irrational ratios were based on nothing, except the relative amounts of oil in a dead whale of each species. Nothing about population numbers, reproductive rates, or how the different whales were withering under the pressures. Nothing about how continuing to hunt the more numerous remaining whales would facilitate total extermination of the rarest. Nothing. Well into the 1960s, the officials of the International Whaling Commission wielded this vapid "unit" like a drunkard waving a loaded gun, willfully miscomprehending the game they were playing.

Here is but a lick of the nonsensicality by which poker-faced men decided, with absolute arbitrariness, on the deaths of thousands of whales — or maybe three times that number; you never know. This is from the 1963 "Chairman's Report":

4,000, 10,000 and 12,000 blue whale units were proposed for the next Antarctic season. . . . Eventually, on the proposal of the Commissioner for Japan and seconded by the Commissioner for the U.S.S.R. a blue whale unit limit of 10,000 was carried. There were 7 votes in favour, 1 against and 5 abstentions. Some of the Commissioners who would have voted for 4,000 units decided not to, as there were three countries who considered their whaling fleet economics could not be supported at such a level of whaling and they might be expected to object within the statutory 90 days if such a proposition were carried. The quota would then revert to 15,000 blue whale units.

Eventually, the asinine "blue whale unit" was dropped. Countries instead allotted themselves whales to kill by species: so many of this kind, so many of these —

Wholly ignorant of clans, family structure — ignorant of everything about whales except how to kill them — the whaling commissioners "decided" that the North Atlantic sperm whales were one "stock" and the North Pacific were "two stocks." On a map, lines were drawn: kill this many here, this many there. They also drew nine lines in the Southern Hemisphere. Thus divided by their enemy everywhere in the seas, sperm whales were set up for industrial destruction.

In the sailing-whaling era, the worldwide kill of sperm whales had peaked in the 1830s at around 5,000 annually, then declined steadily to around 1,000 a year from 1900 to 1920. Between 1964 — the year the Beatles sang "I Want to Hold Your Hand" on *The Ed Sullivan Show* — and 1974, countries reported killing a total of 267,194 sperm whales.

Official whale-kill records are always underestimates because whalers systematically lie. After World War II, from 1947 to 1973, Soviet whalers reported killing 2,710 humpback whales in the Southern Hemisphere. They *actually* killed 48,702. They reported fewer than one out of every seventeen killed. They killed 3,212 southern right whales and reported: *four.*

In 1969, from the deck of a ship already scheduled to cease operations, Peter Matthiessen saw what mere numbers allotted to nations actually meant when men in the space age hunted whales. Matthiessen departed out of Durban, South Africa, with Captain Torgbjorn Haakestad aboard whale hunter W-29 of the Union Whaling Company. The company was, by then, "dying for want of whales." Matthiessen noted:

> Blue whales, right whales, and humpbacks are disappearing from these seas, and to kill them is forbidden. It is also forbidden to kill finbacks north of 40 degrees south latitude, but both laws, the local whalers say, are ignored by the Japanese and Russians.

The whales' muscle would be used for chicken and pig feed; the blood, guts, and bone would become fertilizer; sperm whales particularly would yield a motor oil additive, ivory teeth, and the amino acid creatine for flavoring manufactured soups. "Nothing is wasted," Matthiessen wrote, "but the whale itself."

The day prior, Union Whaling's six ships had killed fifteen sperm whales. But there were some left, and before breakfast the fleet's spotter plane was turning circles over two groups a couple of miles apart, traveling in the same direction. Through the radio, the ships received the whales' distant locations and headings. In 1965 this company killed three thousand; the next year a third that, and this being their penultimate year, 1969, they were determined to get what they could of what remained, before they spent more on fuel motoring out sixty-odd miles to the whales than they could gain selling their oil and carcasses.

Directed into place by the pilot, the ship begins searching where the pilot cannot see: far beneath the waves. The ship's sonar reveals a whale swimming at a depth of a thousand feet. They will follow until the whale must breathe. Not much patience is required. Though whales usually hold their breath for an hour, "this time," Matthiessen learns, "is rapidly decreased by panic when the animal is pursued." He intuits that the whales are communicating their alarm. The mate tells him, "With us they stand no chance."

Over the radio, the pilot and several ships' captains agree on who will take which whales. By edict of the International Whaling Commission, they are not to kill sperm whales under thirty-five feet. Of course, so difficult is this to judge that a two-foot indulgence has been graciously granted; they've thought of everything. Under thirty-three feet, whalers risked a potential fine, per whale, of thirty British pounds. A dead whale was worth about ten times the amount of the fine.

Several whales surface together. There is the burst of the harpoon cannon. Three seconds after the four-foot, 185-pound harpoon hits the whale, its grenade explodes inside her. Then the whole ship shudders as the stricken creature, surging away, jerks the harpoon line taut and a huge spring belowdecks absorbs the jolting of her thrashes. The whale rolls and a well of her hot blood rises, spreading over the cooler seawater. She is still thrashing. The men fire another grenade. Then she is gone, and her body is theirs.

"Already the bright stain on the bright sea is huge and thick, as if it would never wash away," Matthiessen says. "The blood spurting from the wounds is a deep mammalian red, but on the surface of the sea it turns red-red, as vivid as a dye, and the amount of it is awful."

With a grin, the radio operator acknowledges that she is undersized. In eleven minutes from the time of the cannon shot, they are hunting again. Captain Haakestad tells Matthiessen that when the company closes down, the following year, he and the mate plan to start a retail shop in the suburbs. He says that he has never liked the sea, and will not miss it.

169

In 1972, a Soviet whale scientist named A. A. Berzin warned that "absence of stringent international restrictions may prove fatal for the North Pacific sperm whales." That year, the United Nations called for a cessation of whale killing, the United States' brand-new Marine Mammal Protection Act finally shuttered the last U.S. whaling station, and at the International Whaling Commission, the United States introduced the first proposal for a global whale-hunting moratorium.

Which was rejected. The Soviets and Japanese voted themselves increased kill quotas. In 1974, the world's whale-hunting countries reported killing 21,217 sperm whales.

By that time, the extinction trajectory was igniting loudening international opposition. Countries that had quit whaling began advocating restraint.

Unfamiliar with restraint, in 1977, the International Whaling Commission voted to *increase* their sperm whale quota from a misleadingly precise-sounding 12,676 to 13,037. Commission scientists recommended a kill of 763 North Pacific sperm whales. The commissioners set the kill at 6,444, more than eight times as high. That year, at a special meeting of the commission, the world's experts could not come up with an estimate of how many sperm whales remained. That meant that those precise-sounding quota numbers were — in nontechnical terms — total garbage.

In 1979, the International Whaling Commission finally summoned the votes for a moratorium on most industrial whaling, a result so surprising that it stunned even the most ardent whale protectionists.

One month later, the Soviets said they wanted to kill 1,508 male sperm whales. Denied permission, that year they reported killing an accurate-sounding 201. Who knows how many they really killed.

In 1982, the commission passed a resolution declaring that by 1986 the quota for all species would be zero for a decade. Most countries that were still killing whales soon quit. But it was only a resolution; it's voluntary, and there are loopholes.

Killing whales in industrial mode, and too often lying about it, continues to this day, mainly courtesy of Japan, Iceland, and Norway. Iceland is killing thousands of minke and fin whales. When scientists analyzed the DNA of seven hundred pieces of whale meat bought in Japanese and Korean food markets and labeled as being from relatively abundant minke whales, they found that the flesh had actually been cut from: sperm whales, beaked whales, killer whales, dolphins, humpback and fin whales, sheep, and horses. Norway

171

labeled a big shipment of whale meat to Japan: "shrimp."

Japan's ships kill hundreds of whales annually, of various species. And in international court, Japan has been convicted of wrongfully saying that it was killing for "research" while doing no science — a conviction its officials shrugged off, announcing their desire to kill more whales, especially more of the world's great singers, humpback whales.

In 1982, when the International Whaling Commission voted to suspend commercial whaling, thirty-seven countries were members. It's now about ninety countries, as pro- and anti-whaling nations have attempted to stuff this commission's shell with like-mindless votes in exchange for money, as well as other favors related to — money. More than half of these member countries have no history of whaling; many have no coastline. Proposals to kill more whales pop up every year. The only thing about the commission that won't die: a compulsion to kill.

The pioneering whale researcher Roger Payne has said, "It's as if intelligent aliens arrived from outer space and because we couldn't understand their language, we cooked and ate them."

In the North Atlantic, Faroe Islanders stage an annual pilot whale drive, which ends in slaughter; then they cook and eat the whales. At Taiji, in Japan, hunters annually catch and kill several hundred bottlenose, striped, and Risso's dolphins, Dall's porpoises, and pilot whales. (After scaring the dolphins into the shallows, the people corral them and tie them together in bunches by their tails, hitch them to small boats, and drag them backward to where they'll kill them. While being dragged, the dolphins have a hard time getting their heads above water, and some drown. It's a "method" with all the humanity of tying someone to a truck bumper.)

Japan's slaughter guidelines for livestock require that a creature must be rendered unconsciousness before being killed by methods "proven to minimize, as much as possible, any agony to the animal." But livestock guidelines do not apply to whale and dolphin killing, governed by Japan's Fisheries Agency, which treats dolphins and whales as nothing more than blubber with blowholes. Japan's agency people say that dolphins and whales compete with people for food, and so we must — as if in self-defense — kill them, and eat them. Voicing this justification, Japanese catchers sell dolphin meat and peddle live young dolphins to marine parks and swim-with-the-dolphins outfits in Japan and other countries. The real and only reason is money.

Actually, whales help keep the ocean alive. So to kill them is to accelerate killing the ocean itself. To leave whales alone is to help ensure that the ocean will produce more of

173

the fishes and squids that people like to eat. Sperm whales dive so deeply that they bring back what had been long lost to surface waters. Often they poop immediately before diving, delivering to the surface their deeply acquired iron and other nutrients, returning them to sunlit waters, where they can nourish and grow the drifting cells of plankton that require nutrients and sunlight *and* that soak up carbon dioxide to create the green pastures of the sea, forming the very first rung of the entire ocean food ladder. All the flying fish we've been seeing, for instance, as well as all the fish that chase them and all the birds that eat them, are receiving into their bodies some of the molecular matter raised from the deep eternal night by the superpowers of Leviathan.

And as for whales and dolphins being a threat to humanity, one must ask who is really threatening whom, and whether humans alone have lives and families that matter. Hal Whitehead has written, "When you compare relative brain size, levels of self-awareness, sociality, or the importance of culture, cetaceans come out . . . between apes and humans. They fit the philosophical definition of personhood."

Leviathan, cachalot, sperm whale: these are just labels covering a gap where the living world confronts the imperfect evolutionary work in progress known as human empathy.

In truth, there can be no appellation large enough to fit. Needing no reference to human language, this creature — with the largest head housing the largest brain, interrogating the world with the loudest inquiries — lives its out-of-sight existence in the completeness of itself, distant from all shores and far beyond words, carrying on life in the fond company of family members and comrades. That was true long before any first human had spoken a word, and it remains true to this minute.

To destroy a whale is a monumental denial of life and merely one symbol of the human species' rather recent working hatred for the world. We have named one whale "killer." But that shoe best fits the species who possesses feet to wear it. Whales are more appreciated than ever, and whale watching yields more lucre than killing does. In that appeal to our self-centeredness lies their best hope. But even if all humanity gains the emotional and intellectual maturity to finally divest ourselves from harming them, we may yet lose them in an ocean of plastics, chemicals, fishing tangles, spinning propellers, speeding hulls, and noise. All whales now have trouble competing with the aquatic primate for the fishes of the seas. The more humans fill the world, the more we empty it.

Of course, we also put things into the sea. Roger Payne, the scientist who brought humpback whale songs to the human world's attention, and his colleagues sailed for years collecting samples from whales all over the planetary ocean because he was haunted by questions of whether and to what extent persistent toxic chemicals might be building up in their bodies, harming them. In 2019 on National Public Radio's *Living on Earth,* Payne explained that pesticides and other toxic chemicals run off the land, and "what happens is that they end up in the ocean water, hugely diluted." But at the very first step in the food ladder, the tiny single-celled drifters called diatoms start absorbing these substances. Then tiny animals graze on the diatoms, absorbing all they had absorbed. Little fish and squid eat the tiny animals. Bigger fishes and squids eat them. And the biggest fishes, like tunas and swordfish, and the mammals, like dolphins and whales, get all of the toxic brew, concentrated, because, as Roger explained, "fats can hold lots of toxins, so as toxins get into fats, they build into very high concentrations." Whales get bigger than big fish and live longer, and they are unusually poor at ridding their bodies of these chemicals, which consequently concentrate much more than they do in other mammals. Highly toxic polychlorinated biphenyls (PCBs) were widely used in electrical equipment and remain exceptionally persistent and widespread in natural and living systems.

Pesticides continually wash off farms and gardens. Mercury — a common impurity in coal — gets into fishes after it falls from the smoke plumes of power plants. In devising the systems that light, feed, and heat us, we have toxified our lives and the lives of others. Yet, we still have whales.

What Matthiessen saw and learned about the industrialized whaling of the late 1960s made him pessimistic about the whales' prospects.

The remnant bands of the great whales are destroyed wherever encountered, and doubtless the lesser whales will hold out long enough to make it certain that the last of the leviathans will be exterminated along the way. Already the blue whale is practically extinct, and the right whales and the humpbacks are close behind.

From the time Peter Matthiessen wrote those words, another decade and a half ticked by before most commercial whaling ended. But most commercial whaling did end. And Peter got to see three decades of slowly increasing humpback and fin whales, and the blue whale strengthening its slender hold on continued existence. The blue whale had always eluded Peter; it was high on his bucket list. At his request, I helped facilitate a successful encounter off the California coast, where the mighty Matthiessen, who always seemed bigger than life, finally met the biggest creature that has ever lived. The occasion was fitting and overdue. It would have cheered him to see the increasing number of humpback whales that have begun camping just outside our summer surf in easy sight of the beach, pursuing schools of fish that are themselves finally being allowed to recover.

So, was Matthiessen wrong? And Melville, and Lacépède before him? Not in their time. Perhaps their warnings dug in just in time. Maybe they helped save the whales from us. Who will save us, even a little bit, from ourselves?

Whales were the first oil wells; harpoons, the first drills. Oil was energy to light the darkness and to run machines. In Melville's time, another source of energy derived from the ownership some men claimed over other human beings. Now petroleum companies lubricate governments and demand oceans as their own. Energy still comes to us freighted with brutal trespass. For it, we change the heat balance of Earth and acidify the seas, inundate the shores and intensify storms. Energy is always a moral matter. It has tended to reward immoral behavior. It would be ironic if the whales won freedom and we did ourselves in. For the moment, that remains a write-your-own-ending kind of story. Getting hunted to near extinction may have erased much whale culture. It is a distressing story that can be hard to hear. But there is another reason we must not look away: things have gotten much better. Understanding how whales got through the acute times of the past can help us get them through the chronic times of the present.

FAMILIES:
TEN

There is no folly of the beasts of the earth which is not infinitely outdone by the madness of men.

— Herman Melville, *Moby-Dick*

For six days we have seen no whales. Today we will try again, searching for their families to the best human abilities of our eyes and ears and minds and hearts.

"This is the worst stretch for eight years," Shane is lamenting. The year he's referring to — they went *eighteen* days without seeing whales — ships were using extremely loud air-gun blasts to detect potential oil deposits deep under the seafloor. Surveyors can generate tens of millions of such 260 decibel sound blasts. These whales, so extremely evolved to listen, so dependent on hearing the sound-shadow of a squid in the dark, likely cannot cope with such noise.

Being reminded of such "seismic surveys" colors my outlook one shade darker than today's ocean blue. Mood: indigo. When

180

Abraham Lincoln was president of the United States, oil seeped out of the ground in places like Pennsylvania. Now that we've used up the easy oil, times have changed. Humans used to kill whales for their oil; these days, we kill whales just by looking for the remaining petroleum. Seventy-five ocean scientists have written a letter in opposition to new seismic surveys off the U.S. East Coast. It reads in part:

For blue and other endangered great whales, such surveys have been shown to disrupt activities essential to foraging and reproduction over vast ocean areas. Seismic surveys have been shown to displace commercial species of fish, with the effect in some fisheries of dramatically depressing catch rates. Airguns can also cause mortality in fish eggs and larvae, induce hearing loss and physiological stress, interfere with adult breeding calls, and degrade antipredator response, raising concerns about potentially massive impacts on fish populations. In some species of invertebrates, such as scallops, airgun shots and other low-frequency noises have been shown to interfere with larval or embryonic development. And threatened and endangered sea turtles . . . have their most sensitive hearing in the same low frequencies in which most airgun energy is concentrated.

181

To lighten our mood, I tell Shane that, unlike what we've done for the last few days, he should stop listening to whale-free water. I tell him, "Just find whales on the first drop."

We get to the first listening spot and lower the hydrophone and —

He's listening. He points. Over to the southwest.

He hands me the headphones. Loud and clear, three or four whales are clicking through the black depth. Wow, I say, nice. Why didn't we just do that for the last six days?

So we go southwest about a mile. Shane listens again and says, "It's gone quiet there." They must be coming up.

We scrutinize the wrinkled sea for a white puff in the million-mirror glare, looking, looking.

Shane's student collaborator Fabien Vivier, who is French, says softly, "Zere's a blow."

He estimates a kilometer. But none of us —

"How sure are you that you saw a blow?" Shane inquires.

"Hundred percent."

We head that way.

No whale.

But when we stop to listen again, the headphones reveal whales in several directions. Shane hears two ahead of us. Three or four inshore. He adds, mysteriously, "Too many whales for my dreams."

But now he says he hopes we're not being fooled by false killer whales.

What?

"They go, *Buzz-click*. But with enough distance, you don't hear the buzz, just the click."

Could we really have lost the sperm whales we'd clearly heard on the first drop because we let a bunch of *Pseudorca* draw us off the track?

"It wouldn't be the first time."

But then: "There!"

Definitely a sperm whale blow.

"Okay," Shane requests of Dave, "let's poco-poco ahead."

And as we do, Dave also shouts "Blow!" as another whale pops up nearby.

The question: Who? Shane says he's hoping it's someone we haven't yet seen this season. "Her dorsal is really big."

Saying a sperm whale's dorsal is "big" is like saying a newborn whale is "small": it's all relative. Sperm whales have rather tiny dorsals. Anyway, her dorsal is a clue. But when she flukes up, Shane says, "Who the hell is *that*?"

Fabien calls a third "Blow!" He says it's very distant. This time, no one doubts him.

After being underwater for the better part of an hour, a whale will always take that first breath suddenly and forcefully. But this one is so far from us — perhaps two kilometers

— that I can barely see the drifting vapor.

The two closer whales up ahead are now at the surface, about a quarter mile from each other.

"Let's go poco-poco toward the closer one."

Through the silvered sea, we slowly motor. The morning sun skips stones of light across the waters, turning wavelets into a million points of bright. Up from this shimmer, into the low haze, come occasional angular puffs of whales breathing. We're doing science wrapped in a tropical dream.

One small fishing boat, looking lonely in its transit, slows. Two fishermen. One waves both hands over his head, as if he's giving a distress signal. But no: he is pointing to the second whale. We wave back, acknowledging that we've seen both.

Shane smiles and says, "I love that those guys did that for us."

The closer whale ghosts.

We go poco-poco to the fishermen's whale. For a couple more minutes, the whale continues breathing out used gas, packing new oxygen into the freight holds of her blood.

She flexes and her back rises. She flows forward, nods downward, and her flukes rise high, pouring water into the sea. A very distinct half-moon is missing from the edge of one fluke.

Shane says, "That last whale, the left fluke seems super familiar."

So — who?

We seem to be hearing eight whales here. Shane is trying to determine their identity from the flukes we've just photographed.

Shane is entirely present whenever we're at sea. He must be sure that his coworkers note the things that must be noted, and that they are recording the sounds of whales when he needs their sounds recorded. That his drone is ready when there's an opportunity for aerial shots, to index the whales' body conditions. That the suction tags are prepped in case the sea slicks down and we get a chance at a close approach. That everyone's in a good mood and there's enough water and snacks. So he is scientist and business manager and mentor, and he's always paying attention.

But there are moments. Quiet minutes while transiting from listening spot to listening spot. Days without whales. And in the moments when he does not have to play manager, when there are no whales within sight or sound of us, then something haunts him. It's the possibility that the whales he's known will stop returning. The possibility that his sea of sounds will fall silent. That these families, each unique, will begin blinking out. That his life's work will become not a bridge to deep understanding but a walk off a plank into a suffocatingly empty ocean. That it might — end.

In Shane's data tallying births and deaths,

nothing suggests that his concerns are un-
founded.

"Well," chirps Fabien, "three whales in ten
minutes. Good day so far!"

Shane squints into the chop. "It could be
— Unit P. We need to hear their codas."

The smaller clan, which includes the P and
K Families and just a handful of others, uses
an identifying coda having five evenly spaced,
slowly paced clicks. They don't make the
"One, two, cha-cha-chá" coda of the larger
clan in which *all* the other local families claim
membership.

We move a short distance. The directional
hydrophone goes down. The headphones go
on.

Now they're all down foraging, beaming
their pulsed sonar inquiries, sending their
steady "One and two and . . ." clicks throughout the enveloping ocean.

"Oooh — buzzing. She's just chased something!" Shane turns the pole and points, saying, "Most of them are over that way."

I try to picture what is going on thousands
of feet below and perhaps a mile distant.
Once, in a close encounter with bottlenose
dolphins in the Bahamas, I got a glimpse in
miniature of life for a whale with sonar. With
mask and snorkel, researcher Denise Herzing
and I had slipped into clear water thirty feet
deep as four bottlenose dolphins foraged for
fish that were completely buried in the sand.

186

Waving their heads back and forth just above the sediment, the dolphins seemed to mine-sweep the seafloor. I could easily hear their sonar buzzing. A couple of times a minute, a snout would suddenly plunge into the sand and the dolphin would pin and scarf down a thoroughly hidden fish whose hiding hadn't saved it. I was watching the X-ray-like power of their sonic search engine, the most elite hunting weapon in the sea. As they moved leisurely along, reaping so hidden a harvest, I was astonished by their superpower.

Shane had said that most of the whales were over in this direction. Now that we're here, I don't hear a thing. But an hour has passed since they went down, so they're probably traveling upward.

"Okay, let's look for blows."

Minutes pass.

Then Captain Dave announces, "There! She blows!"

Her size impresses even Shane. "Imagine the changes a whale her size has seen," he muses.

After a few minutes, blood recharged with oxygen, she hunches and flukes up — and is gone.

"I know this whale," Shane insists. He thinks a moment, running a whale-tail catalog though his mind. "I'm just not sure yet."

Meanwhile, turning the hydrophone pole

reveals three whales far, far down. With his arms Shane indicates their positions, spanning a forty-five-degree spread. They are tunneling through the ocean in three dimensions while we are stuck up here, stranded in Flatland.

Moving toward them takes us straight into the chop, and spray flies as we pound into each wave, until verily I am drenched. But the water is warm enough, and under this punishing sun, it feels good.

Three brown boobies and two storm-petrels fly by.

Two whales rise together up ahead and raise their breathy flags. They are resting side by side.

Shane points out the most basic, most significant aspect of what we're seeing: "In the vast ocean, to be beside each other means something pretty amazing." Out of nearly infinite space in three dimensions and the fourth dimension of time, all their years of swimming — "they've *chosen* to come right alongside each other." And so here's the thing: such a decision requires effort and persistence; it requires conscious intentionality, consistently applied, diligently maintained, with feeling.

When these two fluke up and slide downward, they begin making codas.

Shane shorthands, "One plus one plus three." One, two, cha-cha-chá.

So, these whales are of the larger clan. But that eliminates only the few families that make up the smaller clan. These could be any of the usual suspects in the whole larger clan, a dozen and a half families.

At 11:05, the first whale we'd seen this morning is up again. She rolls on her side and cuts a little semicircle, and when the tip of her fluke breaks the surface, we can see the scoop-shaped bite mark we'd noticed earlier. But we still haven't identified her.

Shane thinks the semicircular maneuver indicates that she's listening to whales who are below, that she's oriented to them.

She arches into a shallow dive — no flukes out — and makes a run for about two hundred yards. And when she again pops to the surface, another whale is with her, alongside. So Shane was right.

Captain Dave says, "She was just findin' her family."

"Who *are* these whales?" Shane demands. It's driving him nuts.

After a few minutes, they fluke high and dive deep. They are graceful and professional, as though they've perfected a form of Zen in the art of living.

By looking at the photos we've taken, Shane ascertains, within a little while, that we've spent our morning shadowing the L Family.

Mystery resolved.

An hour later, mysteries compound when Shane thinks he hears *another family* to the north.

I can make out the clicks of these new whales, but only faintly; I must concentrate to pull the clicks through the sounds of waves splashing, the high thin whistles of dolphins, the underdrone of ship engines, and the whine of fishing skiffs.

And so we leave the family we've been tracking and attempt to locate and identify the newly heard whales.

And now, again, the ocean seems very large. Large enough to produce different traveling groups of whales who can choose to seek out or avoid each other. And the ocean also seems small. Small enough that whales who travel thousands of miles can be here together and know who is who. And so the ocean — which can strike a person as so bleak and empty and hostile — again reveals itself as profoundly alive.

Just a few minutes later, a whale rises and blows and breathes, smoking her hourly peace pipe.

We poco-poco closer.

It's actually two. A mother and a large youngster of perhaps five or six years. They are touching-close, and breathing in sync.

To one another, they mean everything. And when they dive, Mom flukes up but the younger one just hunches up and vanishes down.

Shane's got the headphones on, monitoring their communiqués. "Mom's talking to her baby."

I hear the coda clacks stop. The sonar clicks begin as Mom goes into foraging mode.

The phrase "talking to her baby" begins orbiting the inside of my head. I wonder what the whales said to one another when they were being harpooned throughout every ocean, when days that had dawned on a family ended with a few traumatized survivors and doomed orphans. It's the same with elephants now. Giraffes. Lions, rhinos, orangutans —. We the people are affecting the conditions by which almost every creature on the planet can live, the rate at which too many die too fast, as many species are sinking. Shrinking forests, melting ice, plowed grasslands, raging fires, drying rivers, and dying corals — diminishment of all the major habitats, proxy for all who live therein, means that the numbers of free-living animals are the lowest ever, and mostly falling, across the board. It means something acutely awful, I think: that the human species has made itself incompatible with the rest of Life on Earth.

On the planet of whales who are talking to their babies, we could have made for ourselves and everyone else in the world a much

191

better deal. We are the only species that makes global problems. It would be useful if there existed a species that solved them. Whales recognize differences among themselves because groups have specialized differently in answering the question "How best can we live where we are?" Why aren't we asking ourselves that question?

And what would it say of us if they all go? Would we miss them? Or, my worst fear: that for most people, their disappearance from the world would be less noticeable than turning off a light. People notice when the lights go off. But if whales disappear — ?

And yet there is life and some time left. And the many people who care. And because of a few who insisted, there are still these whales and a few wild horizons. Life has always been about thin margins and long odds. Perhaps the struggle to keep it must always be that way, too.

Shane looks across the ocean. "It's always exciting to see whales you don't know," he says, sounding as though there's something he's not saying.

A couple of moments later, he adds, "But it's also disappointing. I really wanted these to be Fingers and Digit and the Group of Seven. But it's not them." That was the family he'd spent the most time with, and they became the best-studied sperm whales in the world. Shane falls silent for a few moments. "This is shaping up to be the first year we haven't seen the Group of Seven," Shane says. "No Pinchy. No Fingers. It's kinda depressing."

Turning to me, he says, "It's a weird year. Weird weather. Weird whale behavior. And many of the whales we are seeing are whales we haven't seen in a long time. We hadn't seen Unit L in almost a decade. We saw them twice this week. Unit T, not for seven years. I want to ask them, 'What have you been doing for all these years? I graduated and had two kids; where were you?' " Shane wonders out loud whether the arrival of new whales implies the departure, or perhaps the demise, of familiar whales.

They can go fifty miles in a day; in five years they could go pretty much anywhere. They might have been just nearby, off Grenada. Or way across the Atlantic, off Africa. Maybe they're looking for an old, familiar ocean they know how to live in. Or just trying to get their groove back because the weather has been weird.

"When I named these whales, I was hoping I'd know them for my entire working life. I had assumed Thumb and Enigma would be here as long as I kept coming. We would come back looking forward to seeing certain whales. Every year, the most exciting thing is getting to see the whales we'd left. To see, say, the R

Family again, and to see how Rap and Riot and Rita have all done in the year you've been away. Now you brace yourself; you expect not to see them." He glances at me.

I ask which young whales whose births he was here for have died.

"Oh," he sighs. "That's a long, sad list. When we met the Group of Seven, Fingers had her young one, Thumb. When we returned for our second season, we realized that Thumb had died. We know not everyone can survive. Then Enigma was born to Mysterio. I really liked Enigma. We spent a lot of time with her. She and her cousin Tweak would come to our boat while everyone else was down foraging. I really looked forward to seeing her. And then she wasn't there. And breeding females were passing away. Puzzle Piece, who'd been here since the 1990s, died. Quasimodo, Mysterio — gone. The Group of Seven dwindled to three. So in just the family we've spent the most time with: Thumb was born; he died. Enigma was born and died; I took her death poorly. Tweak died. And Digit is pulling around stuff she got tangled in. That's four babies in a family of five adults. That's devastating."

Shane puts their loss and his grief into context by saying, "No one had ever followed the fate of individual sperm whale families for a decade. We did that. When we looked at the data, we realized that twelve of our sixteen most commonly seen families have gotten smaller. One of every three newborns aren't living to their first birthday. Adults are aging out. These families —. They're dying."

Every whale counts not just as one whale, but as a vessel of knowledge, Shane reminds me. "When young whales die, their grandmothers' wisdom is at risk of vanishing." On this he is adamant: "Each whale has a unique spot in the social network. If Pinchy disappears, you can't replace her with Fingers." Each sperm whale is somebody to other individuals. Their death matters to those who survive them. Relationships create an added layer to life. And they add a layer to the meaning of death.

So what's at stake is not just numbers. It's not just population or diversity or even "culture." What's at stake is: ways of being. What ancient memory banks are being purged, what rich files of life's library are being erased? What's at stake is communities of individuals who know who they are in the world because they know one another. Each whale is a node in a web of relationships. Some play distinct roles within families; some play roles between families. Shane has been explaining that the web of relationships is "greatly hurt by the loss of each node." The daunting question becomes not just "How do we prevent loss of another whale?" but "How do we avoid losing Pinchy?"

The declining trend started in 2009. What has changed? It doesn't seem to be food. Whales poop at the surface before diving. Defecation rates here are twice what they are elsewhere, so they seem to be finding a lot of food, eating well. "They face the realities of being urban whales," Shane says, "of living right next to people." There's pollution, pesticides, cruise ships, cargo shipping, and high-speed ferries. There are plastics to swallow; there's fishing gear to get tangled in. Shane points out, "We're making their day harder. We make it more of a struggle to find food, to stay with their family, to avoid all kinds of trouble that we cause."

In the last three years, four of Shane's familiar whales have gotten tangled in fishing gear. Two were young ones. Turner was a juvenile who got wrapped up in the mooring lines of a fish-attracting device. Turner's mother, Tina, appeared to have tried to help her baby. Not only was it too late — Turner had already drowned — but in the effort to free her child, Tina also got snagged. Then in her violent efforts to free herself, she broke her jaw. Rescuers released her. She vanished.

Shane confides that his analysis shows that these Caribbean sperm whales are declining at about 4 percent per year. If that rate of decline continues, in twelve years the sixteen commonly seen families will each be down to one whale, or be gone. We wouldn't know about the potential loss of this community of whales if it weren't for the work Shane is doing. It makes me wonder about all the other places in the world: which whales may be slipping away?

I have seen living whales bearing the scars and wounds of deep propeller slashes. And more than once I have seen dead whales on the beach who'd been killed by blunt trauma. Cargo and tanker ships are so big now that often no one aboard is aware that the ship has struck a whale. We've come a long way since an enraged whale sank the *Essex.*

In the Philippines, a fifteen-foot-long Cuvier's beaked whale washed up with almost ninety pounds of plastic in his stomach. Their corpses are like bottles all bearing a message: "Your world is killing our world."

Then there's this: A multiyear study of whale stress-hormone levels happened, by chance, to run through 2001. After the 9/11 attacks, while global shipping paused, the whales' level of cortisol, a stress hormone, plummeted. Meaning: routine ship traffic is now intense enough to keep whales permanently stressed. "To me, that's mind-bending," Shane remarks. "It says that what we do to make our lives easier is making their lives more difficult. The whales don't deserve that."

Does Shane think that whales understand

that they are declining, facing a possible crisis?

His answer comes as a koan: "What is the role of traditional knowledge when things go bad?"

He asks me to consider what it means when a group leaves an area they've lived in, as did the Galápagos sperm whales. In the late 1990s in the Galápagos region, sperm whales from the Regular and Plus-One Clans began leaving. By 2000, those groups had entirely relocated a great distance away, off the coast of Chile and far up into the Gulf of California. Perhaps the large group of sperm whales I saw with their newborn in the upper gulf were former Galápagos whales, having come so far.

For the next decade, essentially no sperm whales lived off the Galápagos Islands. In 2011, sperm whales reappeared — a load of them, about 460 individuals. But — they're entirely different individuals. The researchers who documented this changing of the guard published a study titled "Cultural Turnover Among Galápagos Sperm Whales." They found that these hundreds of "new" whales were from two other clans (Short and Four-Plus), which were known across the Pacific, but previously unknown around the Galápagos.

Does this confirm or contradict Shane's belief that whales' culture is an answer to the question "How do we live in this place?" It depends on why the original clans vacated the Galápagos. Those clans had indeed come to different answers about how to live in Galápagos waters. Earlier we discussed how the Regulars stuck closer to shore and meandered, moving shorter distances daily, while the Plus-Ones breezed along in straighter lines farther offshore. Each strategy was an advantage in years having certain ocean conditions and a disadvantage in others. Did each clan leave to track conditions it deemed favorable? As Shane says, "Maybe it's something like 'It's better to go to a place that fits what we know how to do, rather than stay in a place that is changing.' "

Their ocean was changing. The frequency of food-sparse El Niño conditions was increasing, and populations of large Humboldt squid were burgeoning far to the north of where they'd previously thrived. But why, a decade after these sperm whales vacated, would different whales — and only different whales — find the Galápagos livable?

In a mood to press Shane, I ask if the whales really have any way of conferring, of agreeing to leave.

Up against that corner, Shane neither denies nor confirms; he says something more interesting: "Their social lives are too complicated to dismiss that possibility. They have so many relationships at different levels. Remember, one of the reasons cultural groups evolved is to enable cooperation. When animals leave areas, they know, 'This place is no longer good for us.'

"And," he adds, "even to do anything like that — to simply decide to leave — you need a narrative passed along from one generation to the next. You need knowledge of how it was, and what to do when things go bad."

The sea these whales have learned, the sea they are teaching their off-spring to face, may no longer be the sea they live in. As world changes accelerate, will the right answers the whales have learned become wrong? Whales have survived the arrival and proliferation and the onslaught of humans. But will whales' ways begin to fail? We live in a time when those are questions.

"If you lose the sperm whales' library of instructions on how to succeed here," Shane says, "even if immigrants wandered in, they would have to re-figure out how to live in the Caribbean."

Could they? A wolf family in Alaska's Denali area — the Toklat pack — had invented a special way of hunting wild sheep called Dall sheep. They thwarted the sheep's escape tactic of running uphill by strategically moving up high before chasing them in a downslope direction. No other wolves hunted that way. After humans killed the Toklat adult pair and two of their offspring who'd learned the technique, the family's six youngest wolves — who'd been too young to learn sheep hunting from their elders — subsisted on hares; they never hunted sheep. Wrote Jim and Jamie Dutcher in *The Wisdom of Wolves*, "And so that aspect of the pack's culture vanished, erased by a few traps and bullets, perhaps never to be recovered."

But cultures can also be reinvented, at least sometimes. The ecologist Ari Friedlaender tells me that as Antarctic humpback whales recover from the devastating lows caused by intensive hunting, they are relearning bubble-netting in small groups of two or three. Friedlaender and his colleagues are seeing the technique spreading virally there, in real time.

"Human or otherwise," Shane reminds me, "culture is a set of solutions to the problem of how to survive. Lose a family's knowledge of how to succeed as a sperm whale in the Caribbean, and you've made a hole in the canvas. Lose enough families in a clan, you lose the whole picture. You lose the ability of a species to succeed. Lose enough species, and no one will need to colonize Mars. It will be dead enough right here."

The International Union for Conservation of Nature assesses the status of thousands of populations of a wide spectrum of species. About sperm whales, its scientists say:

The Sperm Whale, with a maximum rate of

increase of around 1% per year, is not well adapted to recover from population depletion. Furthermore . . . sperm whales carry high levels of some chemical contaminants, ocean noise is increasing, interactions with fisheries continue to result in sperm whale deaths, and the lingering, socially disruptive effects of whaling may be inhibiting recovery of this highly social species. Some regional populations of sperm whales are declining. . . . The southeastern Pacific sperm whale population . . . has an extremely low reproductive rate (probably below replacement), perhaps because of the social disruption caused by intense whaling. . . . Sperm Whales in the Antarctic showed no substantial or statistically significant increase.

So we seem to be here: there are ship strikes, tanglings, pollution, a changing ocean, competing fisheries, whales choking on plastics. And there's Shane Gero.

Not long ago, Shane got a visit from Sylvia Earle, the famed explorer and defender of the ocean. Around Sylvia, you feel the resonance of a great soul. Sylvia is part scientist, part spiritual leader, part court jester who goes with a wink by the title Her Deepness. Her presence is filled with both humility and well-controlled righteous rage; she has a body smaller than most and a personality just a shade larger than life, luminous and tidal. A hard-charging eighty-plus years young, she seems driven by a passion so pure that it would be incandescent if she were less the elfin sprite she seems.

Anyway, Shane got a visit from Her Deepness and a film crew. They planned to film Sylvia in the water with the whales. And Shane had been feeling stressed about it, because he had built with the whales some sense of what they could expect from him, and he had never invaded the whales' space. (Shane's instincts were not misplaced. A whale named Scar was well known as having a curious, confiding personality. But Shane would later witness what happened when divers on another boat pressed their privilege, kept getting closer than Scar's comfort distance. Scar decided then and there that she'd finally had enough of being around humans. Shane remembers, "That was a sad day.")

When the snorkel-with-whales business started, a few years ago, it relied on the curiosity of one group of whales in particular. It bothered Shane that divers might be taking the whales' attention off their own lives and needs. "When a whale chooses to avoid you, it has taken time to consider you. We don't value time enough. Not other people's time or our own; certainly we don't value how much time wild animals waste trying to tolerate us."

As it happened, with Sylvia Earle they encountered a well-known whale of the Utensils Family. What Can-opener had come to expect from Shane's crew included the boat coming to a respectful distance and putting down the hydrophone. "Can-opener made it into a game. She'd begun waiting about one body length down. And when the boat moved up over her, she'd exhale and rise. Then when we'd lower the hydrophone, she would try gently to mouth it, making the crew play Hide the Hydrophone. And she'd come up and roll slightly as she slowly circled, giving everyone on the boat a good looking over. It's fine being the objective scientist," Shane says, "but to have a whale playing with you, looking you in the eye, you can't resist following around the boat, returning the gaze." What Can-opener had *not* come to expect from Shane's crew was humans dropping into the water.

"So it stressed me terribly," Shane relates, "to have a diving photographer taking Can-opener's valuable time. It felt like a betrayal of what I'd led the whales to expect from us."

Nothing bad occurred that day. After a little time with the sperm whales, the people were all having dinner on the yacht Earle had come on. Shane was talking about his work and the whales. Sylvia was notably quiet.

Later, in a private moment, Sylvia turned to Shane and said, "You feel the burden of the trust these whales have placed in you."

As Shane tells me this, his eyes grow wet. In that one sentence, he says, Sylvia explained to him why he was here. It was something he'd always felt but could never quite put his finger on, could never articulate.

And when he got to shore, he called his wife. She picked up and could hear in his voice that he'd been crying.

And he said, "I finally understand."

And she said, "Tell me what happened."

He told her that Sylvia had pinpointed how, having had this privilege of spending so much time with the whales, he feels an obligation to pay them back. "I owe it to them," he explains to me now, "to be a voice on their behalf. To share their stories. To stand up for their existence. People don't know about their family life, their caring for each other and supporting each other, living with Mom and siblings —. These whales are too important for me to fail." His eyes welling again, he looks away, saying, "So we'll just make it work."

Sometimes things are not as bad as they seem. When Shane first met Unit S, they had three adult females, of whom Samantha (Sam) was the most recognizable. Both of her fluke tips had been cut off. Sally was there, as well as another adult, whom Shane named TBB. And there were two young whales the same age: one with a dorsal fin bent left and the other with the dorsal bent right. Sam the social butterfly would also interact with the J, R, and F Families, as well as others.

"Then," Shane says, "we stopped seeing Sam and Sally. And we stopped seeing the two young whales. TBB joined Unit R."

A few years after she disappeared, Sam stunned everyone by returning. TBB left Unit R and rejoined Sam. Then Sally reappeared.

"What that shows," Shane says, "is that their individual relationships endure separations of great time and huge distances, even long periods with no contact."

How did they remember each other?

"That, we don't know."

Last year, Shane didn't see Sam. Sally and TBB started hanging out with Unit A.

Whales can simply move. Moved whales can simply return.

Shane brightens and notes, "And we were on cloud nine when we returned this year and saw young Jonah alive. And we've got a new one, too: Aurora has lived. So it's not all bad."

The whales we're listening to have changed direction again; they're staying in one small area. The mom who'd been talking to her baby hardly moves, only seven hundred yards from point of dive to her next surfacing blow.

Perhaps she was teaching her child something about diving deep. Perhaps the young one is learning just by her actions that there is a plentitude of squid right there. At any rate, the food appears to be easy right here, right now. I'd love to be able to go and watch them. It's frustrating to be pinned here between gravity and buoyancy when the whales so freely slip to the abyss.

FAMILIES:
ELEVEN

On the boat, we drift and have a bite of lunch. Again, vegetarian sandwiches augmented by copious sugar cookies that have slipped into the lunch pack like stowaways. This has been the hottest day yet under the punishing dazzle of the disk in the sky. It's also the first time we've depleted all the water we brought. If we need more, we'll have to open the emergency box. But the whales have been so generous with their time, it's been well worth a little desiccation.

The sea rolls below, rolls under, rolls past, massaging us in its rise and subsidence, whispering its rumors. By its sheer physicality the sea itself seems alive and companionable beneath a sun that burns its arc as though physics is the mere beginning of it. The slowly throbbing swells bear the rippling breath of a midday breeze. Layers of existence, everywhere.

After some passage of time that I am not marking, a little more than a kilometer and a half from us a whale surfaces and breathes her white-angled puffs. I am no longer startled; I have begun to internalize the rhythms, including the rhythm by which whales, at least, keep their promises.

Of course, our cameras carefully document her high-held flukes as she reinitiates her deep quest.

Shane says, "That's a whale we haven't seen today."

The sea erupts as the young whale we'd seen an hour ago with Mom suddenly leaps clear in two quick, massive sideways flops that detonate white geysers. An upside-down third fling delivers a cratering crash.

Shane wants to see who this young whale's breaches will call forth from the depths. When another whale surfaces less than half a kilometer away, the juvenile swims over to join her.

So, again, three whales: the one who went down a few minutes ago, the young breacher, and now this adult who's just joined the young one, all of them held inside each other's orbits by the stickiness of their emotional bonds.

The juvenile and the larger whale are touching-close. The younger one starts diving to the larger whale's teats, proving that this is Mom — whoever Mom is. So far today, we seem to have seen — I think — eight whales. But we don't yet know who many of these individuals are.

After this little while of sharing the sea air with us, they hunch into their dives. Mom lifts her flukes with high-flying grace. The youngster just humps down, another shallow dive, as if trying to imitate Mom, but — maybe someday.

These whales are headed windward. The wind, however, has put a steep chop over an already challenging swell, so we won't pursue them into that beat-up. Instead we go north for five kilometers. We drop in to listen, and hear whales here, too! Sometimes you go day after day after day without coming upon a click. Some days you hear clicks everywhere you go.

"It's whale soup out here today!" Shane exclaims.

Less than a kilometer away, two whales surface close together. About the same distance northeast of them, a third whale is up. "This makes ten adults and a juvenile so far today," Shane reports.

I dutifully note this. I can't really count them or keep track, because I am seldom certain whether or not a whale I am seeing is one we have already seen.

All the up-whales become down-whales within moments of each other. I do my best to notice their flukes' many individualized marks and scars, and to get useful photos.

"Beautiful," Shane declares. "I still don't

know who these whales are, but they have sexy flukes."

For a little while, I am where I am best, among living manifestations of greater powers that long preceded me and may long outlast us all. For the duration of the encounter, the beauties and truths of them overwhelm the heartache, cleanse everything. For a short interval of time, they have tapped me awake and I feel at home in the world.

Later, not far away, two whales we'd seen earlier surface nearly a kilometer apart. One begins traveling toward the other, then stops. Possible cause: two whale-watching boats and a snorkel-with-whales boat are closing in.

One boat drops several snorkelers in the path of the whale who has put herself on pause. She considers. She alters course. For leviathans once feared as swallowers of humans, they are remarkably shy.

Back when it was anyone's guess whether sperm whales would devour a human like a large squid, Hal Whitehead might have been the first human to intentionally ease into the water among a free group of them. Whitehead wrote, "The whales were hanging like living monuments. . . . I approached them slowly. . . . Two whales glided toward me. As they approached one extended a flipper, gently touching his companion — for reassurance?"

Turns out, sperm whales seldom seek proximity to humans. "Scar would," Shane notes. "But that was his personality. A very social whale."

As individuals with personalities, whales do what they decide to do. "Sometimes I think the young ones get bored when adults want to rest," Shane observes. "When Jonah was very young and adults were resting, you'd see her going, basically, 'Wake up, wake up.' Then she'd come over to the boat like, 'What are you guys doing?'

"The whales consider themselves different individuals," he adds. "So we have to treat them that way."

Think of tribes. Tribes of other beings. Other beings with other minds living other lives on the same planet. Different, certainly. But fundamentally, not really very different. They mean something to one another, and so their lives mean something to them.

Perhaps that should mean something to us.

A large, dark whale surfaces. And a midsized juvenile too, not far away. Between shallow dives they talk. Soon they're resting together like boats tied up alongside each other, having a gam. It's easy imagining this mother looking at her baby, her child looking back, reinforcing with silent gazes the bonds they affirm hourly with their proximity and reavow with their codas of arrival and departure,

212

codas passed like keys through generations in each family, whale to growing whale.

We slowly eliminate the short distance between us and these two whales, until we are directly alongside them, inhaling their breath, almost close enough to touch their wrinkled wetness.

They submerge together. But they remain so shallow that we can clearly see them right there.

They rise directly under the tip of the twenty-foot pole Shane is holding. This is as close to old-time whaling as anyone evermore should be, and the proximity is thrilling.

Poised at the tip of the pole is not a sharpened piece of iron. Instead it is a question on a stick. The question: Who are you? Shane's implement is a hand-sized suction-cupped electronic device that will create 3-D visualizations of the next six hours of the whale's movements through deep ocean. A glimpse, nothing more, but a glimpse of something never before known.

When the adult again breathes and her back is shedding water, Shane taps the tag onto her skin. She barely flinches and does not change direction or pace. It must feel like a remora, those hitchhiking fish that suction a ride, a familiar sensation. After a short, shallow submerging they're back up, yet again, and we see that the tag's four suction cups are holding nicely. The ten-thousand-dollar device is timed to detach at six p.m. today; it will float and send us its location. In the morning, we will pick it up.

Ten minutes later, they go down. Three different individuals come up.

The hydrophone reports whales continuing to click in from various surrounding directions. All day yesterday, we followed whales going south. All day today, they are going north. Between dives, the distances add up. They've taken us far from where we first encountered them.

The blue-gray sea is slick and hazy-bright. It is both eternal and instantaneous, and the whales that this ocean has brought forth seem, in their pacing and their scale, to reflect the enormousness of all things past and present. Something like time must be passing, but I feel suspended in an infinite moment that almost vibrates in place. Perhaps from the whales I have learned something about living.

One whale stops hunting. Over the next few minutes, others also silence their sonars. Up. Up. Up.

A few minutes later, the dark heads and backs of two whales shatter the harsh sparkle of the sea like tiny newborn islands, the coastline of their bodies generating their own surf. Their white puffs drift on the distance.

Three others burst the surface. So, a total

of five up now. One is our tagged lady.

When they all dive, I put on the headphones and am amazed at how *loud* they are, clicking out their clacking coda-codes of recognition, braiding their messages of bonding and belonging. Their sound, highly styled, is percussive and precise, like castanets. As I listen, the codas go in and out of phase with each other. Sometimes they're perfectly separated. Sometimes they completely overlap, like conversations at a busy table.

For a couple of minutes I listen to their affirmations, these repeated code words. And then the rally subsides. They switch to the business of searching. Their echolocation clicks begin coming into my ears. *Tick. Tick. Tick . . .*

With time now for a break, most of our crew takes the opportunity to jump into the water. "Put your head under — you can hear the whales." Vibrating an enormous sphere of water around them, the whales sound-saturate the ocean as they level off perhaps a thousand yards down. Journalist James Nestor got eye to eye with sperm whales in the Indian Ocean and noted: "I heard a thunderous crack, then another, so loud they vibrated my chest. Two sperm whales emerged from the shadows, scanning us to see if we were a threat. Within just a few feet of the mother, the click patterns changed, becoming slower, softer. They sounded to me like the sounds

215

sperm whales use to identify themselves to others in the pod. The whales were probably introducing themselves. They were saying hello."

Half an hour into their present deep foray, they've clearly doubled back. Perhaps they found a dense aggregation of squid and are running the length of it and back, giving them a thorough working over.

For another twenty minutes we drift along in the company of the large juvenile who is waiting while her family quaffs armadas of prey. We, too, wait patiently, trying poco-poco to stay with them. Listening occasionally.

As soon as a whale pops up in the distance and begins purging her lungs, the juvenile we've been hanging with ditches us.

"Joining Mom" is Shane's safe guess. We scan with a tag-locating antenna. Sure enough, that's our lady, the mother who is carrying our tag. She was down fully fifty-nine minutes. Another family member slits the sea surface and greets the baby.

I slowly realize that I'm no longer seeing "sperm whales." I'm seeing a family. And they seem to me, now, a rather humble family indeed. Large in comparison to us, but dwarfed by the greater unknowns of their present and their future.

A kilometer away, two more adults take their inspirations, then burrow deep.

We're about ten kilometers from shore. From various directions, whales toll themselves with steady clicks, ticking up into our headphones. When our tagged lady flukes and dives, she is going west, toward deeper water, where the farthest squid of day will soon be following night's darkness toward the surface.

The smaller whale begins swimming toward three more whales who've just surfaced in the glare path of the lowering sun. Remember how they'd moved only seven hundred yards? In their next dive, they travel more than four times as far. Perhaps they have a plan in mind.

The small one tail-slams the surface as the other two fluke up in rich light and dematerialize. Is this little whale having a little tantrum, frustrated about being unable to follow, or just a frolic and a little fun? At five p.m., two whales burst the surface a kilometer and a half south of where they dived. We've followed whales all day, and they've made one big lap. We'd go poco-poco now, again, but it's time for us to quit this dream and keep our promises ashore. It's getting late for humans who have miles to go and equipment and boats and data to deal with. And dinner to make and the dishes to wash before doing it all again tomorrow.

217

Who knows where the whales will be when we wake at dawn? Do *they* know, these vagabond whales?

I'm having coffee and breakfast while looking past the moored boats. The Caribbean Sea is bluer than blue. Shane Gero and his crew will be here in a few minutes.

Seaward, I don't see water and sky and light clouds. That reality is mere backdrop for the deeper images my mind envisions. My mind sees: whales searching, breaching, and hunting, ticking off their sonar and their handed-down codas. Babies greeting returning adults. Whales living with their families in their miles-deep universe, from which they must journey up to the innermost edge of outer space for every sip of air. A green turtle periscopes the view. I see swallows; and, behind them, pelicans; and, behind the pelicans, frigatebirds; and, beneath them, thousands of feet down in the dark, Jocasta and Laius and many, many whales.

Nearly a year and a half later, I often recall the privilege of peeking into Shane's world — and the whales'. And then out of the blue, I get this e-mail from Shane:

"I thought you'd be happy to know that Jonah is still with us and getting bigger, Digit has gotten free of her rope and will get the chance to grow old and — as the only female offspring — eventually lead her family unit. It seems like the high rate of mortality has slowed."

Yes; I am *happy* to know that.

■ ■ ■ ■

Realm Two:
Creating Beauty
SCARLET MACAWS

■ ■ ■ ■

There is no excellent beauty that hath not some strangeness.
— Sir Francis Bacon

Like flying flames of red and yellow and blue, they transited overhead. From unbroken forest they flew over the wide river, to vast continuing forest rising from the far bank. They are huge birds, improbably beautiful, made by and for a world that, itself, is almost impossibly beautiful. What most struck me: the visibility of their emotional bonds. Macaws were prone to being in pairs; a flock of twelve was clearly six pairs of two, the paired partners flying sometimes so close alongside that they looked like one bird with four wings. I wanted to know more about these seemingly romantic, emotional birds. And so I returned, planning to stay longer, wondering whether such deep bonding and great physical beauty are in some way related.

SCARLET MACAWS

Parrots (Psittaciformes)

New Zealand Parrots (Strigopoidea)

True Parrots (Psittacoidea)

Cockatoos (Cacatuoidea)

Psittacidae

Psittrichasiidae
Indian Ocean island parrots, several species

Psittaculidae
Mostly Asian and Australian, 170+ species

Macaws
Approximately 19 species, including scarlet, red-and-green, and other macaws

New World and African parrots
Approximately 160 species, including Amazons, conures, and African greys

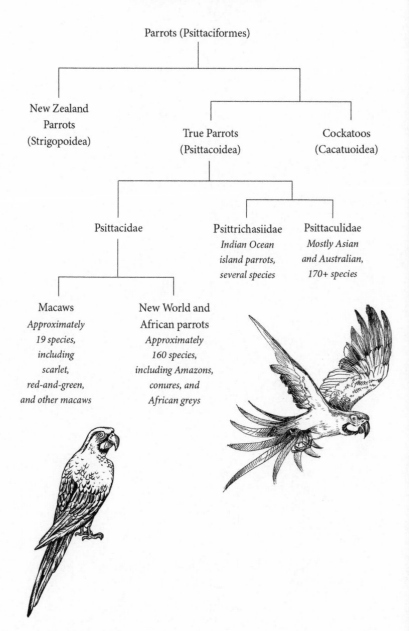

BEAUTY:
ONE

Two scarlet macaws are complicating our attempt to eat breakfast. For two decades in the Peruvian Amazon, Tabasco and Inocencio have straddled the wild and the human-settled worlds. This morning at the exquisite tourist lodge attached to the Tambopata Research Center, they are working the open-air dining area like professionals, transiting from rafters to railings to detect any soft spots in the humans' defenses of pancakes, rice, and rolls.

When Don Brightsmith gets up, he says to me, "Guard my plate, please." At fifty, in his baseball cap with his grizzled beard, Brightsmith is very much the academic field biologist. Along with Gaby Vigo, Don directs present-day research on the free-living macaws in the surrounding rainforest. Gaby is Peruvian; Don was born in the United States. They met fifteen years ago at a conference. At the table with us is their other great collaborative effort: their daughter, five-year-old

Mandylu.

Around twenty-five years ago, over the course of three years, former researchers here at Tambopata rescued about thirty macaw chicks of two species. Each of those chicks had hatched second or third in their nest; their parents weren't feeding them. The researchers hand-raised them. They were dubbed the "Chicos," the kids. They were never confined; they easily rewilded, attracted wild-born mates, and acquired nest sites. They also never forgot their roots, and often returned to the research center to pilfer food.

Even for mere humans, individual macaws can be highly recognizable. Tabasco, twenty-three years old, has a little bald streak on his neck. Through more than twenty molts in his long life, one particular feather always grows back white. He visits the lodge most days. And he is subordinate to Inocencio, who is two years older and also visits most mornings.

Inocencio has one distinctly almond-shaped eye. The blueness of his blues is deep and intense. On Inocencio, the distinctive scarlet macaws' yellow shawl is unusually wide, as if he wears a sunny cape. He's heavy, he's big; he struts like a bad guy. But he has his nurturing side; when he was mated to Chuchuy — pronounced *chew-chewy* — unlike most male macaws, he incubated their eggs.

■ ■ ■ ■

A man named Manolo, who brings our breakfast, energetically chases the Chicos off with a water-filled spray bottle. They know the drill and take it in stride. Manolo has other duties; the Chicos, retired from breeding, have faith in patience. I find fascinating their hybrid personality, wild and tame, harassed and indulged. They have lived long, and prospered.

Chuchuy, twenty-four, arrives. She sports a few green feathers in her scarlet head, and her wings are turquoise rather than the blue of most other scarlet macaws.

I rise and ask Gaby if she'd like anything. As she turns to answer, Inocencio crashes into the plate of little Mandylu. Gaby leaps up, waving her arms. She rolls her eyes and explains, "Inocencio targeted her because she is small, you see? And if it's pancakes? He's insane for pancakes, trying, trying — until he gets one. Tabasco is more hesitant. Inocencio, he just goes for it."

Familiar though they are, the Chicos frighten little Mandylu. They are immense at her scale. And fast. And they violate her sense of propriety. "That was *my* pancake," she asserts, tearfully.

I explain that just as she herself saw it and took it from the plate that was placed before her, he sees it exactly the same way. You wanted a pancake, and you took one; he wanted a pancake, and he took one. To my surprise, Mandylu seems to accept that explanation. "C'mon," I say, "let's get you another one."

I have come specifically for the handful of spectacularly huge, spectacularly hued long-tailed parrots native to Central and South America that are called: macaws. And I am certainly getting: macaws. "Macaw" sounds like one species; there are about a dozen and a half. Largest of all, at more than a yard in length and with a four-foot wingspan, is the hyacinth macaw, which inhabits vast wetlands such as Brazil's Pantanal. Our pirates in the rafters, scarlet macaws, are only slightly smaller.

Macaws are among the more than 350 parrot species with us worldwide. Humans have labeled various other groupings of their order "Amazons," "parakeets," "parrotlets," "lorikeets," "lovebirds," "conures," "cocka-toos," and "cockatiels"; there are several other appellations, including simply "parrot." They are the modern twigs in a brushy evolutionary branching of birds with deep roots. The reptilian lineage that bequeathed birds *and* mammals diverged roughly three hundred million years ago. One lineage evolved through dinosaurs to become birds.

Mammals are not descended from birds; nor are they more evolved than birds. We share ancient ancestors, giving us many similarities; we've been on separate trajectories, giving us many differences. Birds recognizable as parrots have been stirring turbulence into the air for fifty million years. Like all living things, they continue evolving.

Most parrots eat fruits and nuts or seeds. Most will not eat insects — very unusual for birds. But then, many things about parrots are unusual for birds. Including their fondness for our breakfasts.

Gaby says, "If you offer them wild fruits that they can get in the forest, it's like a joke to them. They are like, 'Hey, I can get these any time.' They just toss them. They like bread."

Tabasco crashes Don's place setting and pinches a piece of sweet cake from the basket. No harm done, and Tabasco and I, at least, are fine with this. He sits on the rail and rolls dough balls with his parrot tongue. I'm thinking, "Let them have their doughy pensions."

Don asks, "Have you ever touched a parrot's tongue? They're like soft leather, and they're dry. They're very interesting little organs, and parrots experience a lot of the world through them." At home, my wife and I lived for years with a personality-packed little adopted green-cheeked conure named Rosebud who often tested the food on our plates by touching it with her bill. If it passed bill muster, her next step was an exploratory tongue touch.

Gaby says that Tabasco targets white people's meals. He realizes that if you're white — almost all the tourists are white — you're going to be afraid of him and he can land on your table and snatch food. "Tabasco knows that brown people understand his game."

Many parrots are green, basically. Which makes sense. The big macaws, however, are bright splashes of extravagance, colored like outlandish cornucopias of tropical fruit. Which makes *no* sense. The scarlet macaws look unreasonable. They have scarlet heads; their wings and long, streaming tails are blue and blue-green and scarlet, and their identifying feature is a bright yellow shawl splashed across their shoulders. How could such profligate colors exist? *Why* would birds evolve such beauty? The human in me wonders: Do birds see beauty in their plumage, hear it in their songs? Why does so much that is not intended for us seem so beautiful to us?

Surely song not intended for our ears, allure that is none of our business, cannot seem beautiful to us alone. And if the perception of beauty is shared, well — can it be true that, for all creatures, the whole world brims with beauty? Can it not be?

Here among so many beauties, the questions seem inescapable but the answers elusive. I do not intend to let them get away. But thinking about all this will take time, and may have to be approached indirectly.

Chuchuy's first mate (she was ten years old) was wholly wild, not a Chico. So was Inocencio's. Following divorces with those firsts, Chuchuy and Inocencio paired. Being a year apart in age, they had not been raised together. If they had, they likely would have responded to each other as siblings, and not mated. Chuchuy and Inocencio nested near the lodge. They were driven out of their nest site by a young pair a couple of years ago.

The Chicos helped researchers understand the surprising extent to which individual macaws can have notable personalities and quirks. Ascencio, a member of the species known as "red-and-green macaw," so loved to eat a special sweet Christmas bread called panettone that Don once felt forced to hide his box of panettone under their bed covers before they left the station for a few hours. And — you probably know where this is going — "When we returned everything was torn up. He found it, and shredded the box."

Gaby adds, "They know everything."

Parrots have been called "the humans of the bird world." No offense was meant. It was a reference to genes that give both parrots and humans long lives and clever minds.

231

A parrot's brain and brain stem are large relative to body size. Their brain-to-body ratio is similar to primates'. From the science and from their behavior, it's clear that a parrot's mind functions on par with the monkeys who share this forest.

Scala naturae, "the ladder of life," is an ancient concept, an all-too-human approach to divining some natural hierarchy in all matter and living things. Originated by ancient Greek philosophers (Plato, Aristotle, and others) and later advanced in Christianity, it came to the very convenient (for us) and catastrophic (for the rest of the living world) conclusion that rocks are at the bottom, next come plants, then certain better plants with nicer flowers, lowly animals followed by more elevated animals, and then, atop the ladder: us. This was presumed to be the natural order of things for several thousand years.

The establishment of modern sciences picked up from there, starting in the late 1700s and mid-1800s. Astronomy, aided by telescopes, detected evidence of deep time. Geology showed that Earth and the life on it had once been very different from the world we experience, and that it continually changes. The study of evolution has revealed that living things change in nature similarly to how they change into new varieties under selective breeding by farmers and animal fanciers. These revolutionary ways of investigating existence gave us radically altered notions that forever disrupted our perception of who we are and where we stand. Much that was, and much that is, preceded us. We are neither the center of the universe nor specially created. Needless to say, the discovery that humans are not the center of the universe terrified many people. Many remain terrified.

This leaves us with two problems. One, most people, especially in Western cultures, have inhaled *scala naturae* subconsciously, so embedded is it in our traditions and stories and our cultural disrespect with respect to the world. Many assume that we are the perfect expression that the universe intended. Consequently, we deem the world for us, as ours, believing that bearing no responsibility, we may act on the world with impunity. Two, we assume that the more similar other beings are to us, the better they are. We have a very hard time accepting that the problem-solving intelligence of a raven, a parrot, a dolphin, and an ape — not to mention (but we must) octopuses and certain fishes — are all equivalent to most primates'. We forget that everything alive has made the whole journey with us. Indeed, most have traveled much longer to get here.

Our subconscious imbuement with *scala naturae* is why we are always "surprised" when an elephant rescues her baby or a wolf uses strategy, though they've been doing

233

these things since long before humans existed. We've hardly noticed. The basis for our surprise is our ignorance, our self-isolation, our insecurity, our need to be the best thing that has happened since stars were born. When researchers recently showed that fish known as cleaner wrasses can recognize themselves in a mirror — long considered a proof of self-awareness, thought to be an exclusive attribute of only the biggest-brained elites — scientific journal editors wouldn't publish their study unless they publicly called into question the validity of the mirror test itself. An aware fish doesn't sound threatening, but apparently the fact of that level of intelligence in a fish is more than some humans can bear.

Parrots have been perfecting parrot-ness for the aforementioned fifty million years. That's roughly comparable to the first apes, which were in full swing by about forty million years ago. We might think of a hundred years or a thousand years as a long time. These beings have all been here, continually honing who they are, through spans of times that are essentially impossible for our human minds to fully grasp. It's been a long, strange trip. But here we are, all together now.

Iglecita, another scarlet macaw at Tambopata, is tiny for her species. As a chick, she almost died. For days, she was very weak. One journal entry about her reads, "I hope Iglecita makes it through the night." She is quite particular about letting only specific people preen her. She usually visits the research center only during breeding season, roughly November to March — with a notable exception. Gaby says, "We have one volunteer, Sandra, who has been here four times. Three of those times it was May, far outside the breeding season, and Iglecita came. Somehow, Iglecita appears *every time* Sandra is here. Last time, Sandra hadn't been here for three years. Everybody was joking, 'Iglecita's gonna come.' She came!"

When they have trouble, the Chicos come home for help. When one of them got stung badly by bees, she went to the lodge and stayed in the rafters. When another, Avecita, was badly ill with an internal infection, Gaby recalls, "She arrived so weak that she might have walked here. She was in horrible shape. She spent ten days in the house." She recovered. Don says, "It was another example of them acting on a concept of 'This is a safe place to come when things are going really badly.'"

Parrots are capable of recalling past events, thinking ahead, taking the visual perspective of others, and sometimes creating novel tools to solve problems. These are cognitive feats, "thought until not long ago to be uniquely human," says one group of researchers based at the University of Cambridge. The parrots haven't changed. It's as if we are just waking up from a long journey through space and having a look around an interesting new planet. What scientists considered "hallmarks of human intelligence" before they realized they were also hallmarks of ape intelligence — toolmaking and social strategizing — happen also to be hallmarks of the intelligence of parrots and of birds of the crow family (crows, ravens, jays, rooks, jackdaws, choughs).

Certain birds rival apes as toolmakers. New Caledonia crows make hooked tools, something even chimpanzees don't do. And they make barbed tools from strips of particular palm leaves, with a thicker end to hold and a narrow tip that's effective for getting insects out of crevices. This is a rare example of crafting, where it takes several steps to fashion a tool. Juvenile New Caledonia crows stay with their parents for up to two years, learning toolmaking by watching closely. Crows in different areas across New Caledonia make their tools a bit differently, meaning that the species has culturally transmitted what scientists have shown are "multiple traditions." Researchers noted that in tasks designed specifically to assess ravens' general planning abilities, "the birds performed at least as well as apes and small children in this complex cognitive task." (In the mid-1800s, the Reverend Henry Ward Beecher had reputedly observed, "If men had wings and bore black feathers, few of them would be clever enough to be crows.")

Apes don't have much, if anything, over macaws and ravens. Not in brain-to-body size, social skills, toolmaking, or puzzle solving. Like the great apes, common ravens can follow a human's gaze not only into the distance but also behind visual barriers. In experiments where subjects must use a small stick tool to access a longer one, most gorillas and orangutans did so right away. And so did New Caledonian crows.

In experiments, several macaws and an African grey parrot learned to take a non-edible token instead of food so that they could then trade the token for food they liked better. Thus they understood delayed gratification and the value of currency. The macaws were as good at this as chimpanzees.

The brains of macaws and other parrots, of the crows and their family, and of the great apes evolved differently and are organized differently. The lines leading to mammals split from the reptile lineage tens of millions of years before birds evolved. Peak mammalian intelligence and sociality and peak bird intelligence and sociality evolved independently. They are two separate pinnacles of developed minds. Likely these mammalian and avian pinnacles of intelligence evolved because the creatures needed the computing power required for dealing with complex social groups. With differently organized brain hardware, they've converged on comparable capacities. Similar need, different route, similar outcome. This all happened long before humans existed. But now that we're here, we can consider these other creatures amazing.

In experiments, an African grey parrot named Griffin, trained to name different objects, watched a human researcher deposit into a bucket two different types of items in a three-to-one ratio (for instance, three corks and one piece of paper). The researcher then withdrew one item without letting the parrot see what it was. When asked to identify the removed object, hidden in the researcher's hand, the parrot had to look into the bucket and determine what it was. Griffin tended to be right over a large number of trials with different kinds of items. This is called probabilistic reasoning, and until this recent study, researchers believed it was limited to a few mammals. No one had previously asked a parrot.

Griffin also learned the names of shapes of various three-dimensional objects and was able to apply those names to flat drawings of these shapes. Moreover, when most of the flat shape was hidden, Griffin could often still say what shape the partially exposed drawing represented. This shows that parrots can generalize a concept of shapes of real objects and apply them even to partially hidden drawings.

Like chimpanzees, some species of crows, scrub jays, and ravens change their behavior when a competitor sees them hide food. Jays are more on guard about theft if they themselves have stolen from someone; it takes one to know one. Understanding what another bird might do because you know what you would do is called "experience projection." To see things from another individual's viewpoint, you have to understand that they can understand. You have to realize that they have a mind. Not long ago, most psychologists believed that only humans understand that others have minds. Now some psychologists and other scientists are realizing, systematically and with proof, that we share the world with other kinds of minds.

Reacting to being watched requires another capacity: a concept of time. You must understand that *in the future* the watcher might steal what you've tried to store. Understanding the past and envisioning the future is sometimes called "mental time travel." Young scrub jays learn how to choose the best acorn-storing sites — places of lower humidity where nuts will last longer — from their parents. But they won't use their preferred sites if they have reason to think they are being watched. Yes, I said reason — to think.

Tabasco first bred at age ten, pairing with a wild bird dubbed Señora Tabasco. She used to visit the lodge with him. One of Tabasco's daughters, named Tambo, nests in a human-made nest box near the lodge. She comes to the lodge in the tradition started by her father. And she has introduced her wild mate, Pata, to the ways of the lodge as well. "Pata acts like a Chico," Gaby tells me. "To Pata, everything is new and interesting." One of Chuchuy and Inocencio's wild-reared offspring, ten-year-old Heredero, likewise learned the routine from his parents and visits frequently for a snack. That some of the tameness and food habits of hand-reared macaws gets adopted by fully wild mates and the wild partners of their offspring shows how easily and alertly macaws observe what the other birds know, following their example. One wonders whether some of this information gets communicated directly, as if one bird is saying, "Follow me, act natural, and just do what I do." However they do it, absorbing the local habits of their kind, even when it's unnatural, comes naturally to them.

They flexibly learn particular ways that make the world work for them in the place where they live, acquiring from each other the components of their culture. The Amazon

rainforest is one of the planet's most complicated living systems. What they need — food,
minerals, water, nest sites, mates, allies, and
safety — all occur in different places and at
different times. Their required skills are
multiple. And although their dense forest may
mask their cultural competencies from human view, the things that fully wild macaws
learn from each other must be much more
complex than grabbing dinner rolls and
avoiding waiters with spray bottles. But obviously they can learn that wholly artificial way
of life, too — if it's to their advantage.

Eyeing Tabasco with a mixture of exasperation and admiration, Gaby says, "He's always
calm, never nervous. When Tabasco was
young, he was always looking for new things.
You'd see his face, like, 'Ah! Something new!'
He used to do a lot of exploring — and
destroying." Tabasco is the only one who
regularly went into researchers' bedrooms.
He still does, but not every day as in the past.
"He's no longer as curious," Gaby notes wist-
fully. "Everything is familiar to him now."
Fondly yet matter-of-factly, she adds, "The
Chicos have taught me ninety percent of what
I know about macaw personalities. Having
them around feels like cheating."

Across the animal panoply, individuals differ
in personality. They may tend to be relatively
shy, bold, curious, active, quiet, calm, or

241

nervous. In everything from birds to spiders, personality affects how individuals forage, socialize, explore, perceive danger, and choose mates. Zebra finch females who like to explore prefer males who are also adventurous. Field cricket females prefer bold males. Female parakeets prefer males who've shown intelligence at solving a task. Personality is necessary for innovation and, therefore, ultimately, for culture.

The living world holds much more innovation than you'd think, causing many animals to do seemingly improbable things. I have seen the common blackbirds called grackles wading into shoreline shallows of tidal streams and catching and eating baby eels migrating from the sea. I have even seen skuas sipping milk from the nipples of a nursing elephant seal. And perhaps more bizarrely, in one population of little chickadee-like great tits in Hungary, these birds weighing just over half an ounce search for, kill, and eat hibernating bats. Never say never!

Personality differences help create specialized individuals. (Later in my time in Peru, I will come to a startlingly unexpected realization: that the *ability to specialize* can be a first step on the strange road to the emergence of *beauty*. We'll get to that.) A capacity for specializing is very practical, because adapting to change often requires specialists. Where I live, herring gulls exploit human-caused changes, but they do so differently from one another. From the same breeding colony, some herring gulls habitually travel out to sea to follow fishing boats for discarded catch, while others forage at landfill garbage heaps and yet others continue with traditional natural foraging, hunting for crabs, whelks, and clams. You might call these collective specialties the gull colony's cultural environment.

Along the same lines — and more clearly learned socially, so more clearly cultural — sea otters learn a foraging specialty from their mother and keep it, throughout their lives. Each hunts only a limited selection of all the possible prey in their area. Finding and utilizing each kind of prey — abalone, urchins, snails, mussels, sea stars, and an array of crabs — "probably requires radically different skills," say researchers. So the array of specialized skills might be thought of as otters' cultural milieu. Shorebirds called oystercatchers specialize in either stabbing or hammering open mussels. Chicks whose parents stab mussels develop the stabbing technique; those whose parents hammer mussels, hammer them.

Macaws, like almost all birds, are adult-sized by the time they first fly from the nest. But they take several years to mature and start breeding, because they need time to learn how to live. A lot they learn socially; that's their culture. Knowing what you're doing and how to do it is the great ace in the hole of adult animals. Compared to young individuals, adults are better at getting food and have lower mortality rates from predation. But learning skills takes time, and once skills have been learned, applying them is more efficient than learning a new skill. This again leads to: specialists.

Gaby and Don, with their front-row seat for the Chicos, have been able to see personalities translate into different ways individuals approach living, and how those birds' wild mates and offspring picked up some of their techniques and tricks. Bird researchers have documented specialists among eagles, penguins, albatrosses, cormorants, murres, oystercatchers, various songbirds, gulls, and many others. When I trained hawks, I noticed that individual birds developed particular techniques for particular situations and often became especially skilled at catching certain prey in certain ways. Only one wolf family in Yellowstone specializes in hunting bison; only one in Minnesota specializes in fishing. If an individual's specialization spreads and others come to share the skill or habit, then that skill or practice becomes cultural.

One morning in 1921 in Great Britain, someone opened a door to get the milk that had been delivered and was surprised to discover holes in the foil tops of the bottles. The crime spread, year by year, until, twenty-five years later, milk bottles in roughly thirty towns across the United Kingdom were being pilfered by cream-sipping thieves. The perpetrators: little chickadee-like birds called blue tits. For nearly a century, behavioral ecologists wondered whether each bird had to figure it out on their own or if there was one genius — or maybe several — whose piercing insight had been widely copied and spread through social learning.

To investigate the birds' capacity to spread behavior culturally, researchers caught two wild adult blue tits from each of eight populations, gave them four days' training on how to open a puzzle box stocked with live mealworms, then released them where they had been caught — having already distributed puzzle boxes in their territories. In the neighborhoods where trained birds were released, after three weeks, 75 percent of all the birds knew how to open the puzzle boxes. (In neighborhoods where birds were not trained, it took about two weeks for the first birds to figure it out, and the skill spread slowly and to well under half the birds in the three-week experiment.) Over two generations of blue tits, the skill passed to young birds. And there was a pattern: juvenile females were twice as likely to acquire the new skill as juvenile males or mature females. Least likely of all to learn the new trick: adult males. Across a wide swath of animals, the young, especially young females, appear to be the best learners (probably because young females generally divert less time to squabbling for dominance). As groups learn certain new skills for making a living, cultural quirks and specialties add diversity to the survival tool kits of different populations. This is a boon when change comes to a population. And change always comes.

The world is changing very quickly now, because we are changing it. Enclosed shopping malls, which are now part of human culture, are also part of the culture of urban birds. Pigeons and sparrows have learned to get into malls — sometimes by using motion sensors to open doors — and forage on the floors for crumbs. (I also see them in underground train stations in New York City.) Urban sparrows and finches often bring cigarette butts to their nests; they've somehow discovered that the nicotine kills bugs. (When I raised homing pigeons, it was common practice to buy bags of tobacco stems for nesting material, to control lice.) In some places, crows drop nuts on roads and wait for them to get run over by cars. In at least one place, they do this at intersections, so they can safely walk out and collect their cracked prizes when the light turns red and the cars stop. They've developed answers to the new question "How can we survive here, in this never-before world?"

The point is, individuals vary, so cultures vary. Cultures evolve and respond to change. And that means: cultures can be damaged. Can be lost. When populations plummet, traditions that helped birds and other animals survive and adapt — vanish. The modern world's nine thousand or so bird species contain about eighteen thousand regional variations usually called subspecies. To avoid losing bird biodiversity worldwide, we'd have to ensure the existence not just of the nine thousand species but of all eighteen thousand subspecies — for starters.

I've mentioned that conservation owes much to an important — but still too limited — concept called biodiversity. "Bio" refers to living things, and "biodiversity" is a short-hand reference to the diversity of all life on Earth. This simple term helps us organize our thinking. Ecologists usually think of biodiversity as having three main levels: the "genetic diversity" within any particular species; "species diversity," or the number of species in a given region; and "habitat diversity," referring to the diversity of habitat types, such as forests, grasslands, coral reefs, sea ice, and so on. Biodiversity is usually approached as a gene-pool thing. But does that cover it? Not really. There's a fourth level we are just becoming aware of: cultural diversity. Skills, traditions, and dialects that animals have innovated and passed along culturally are crucial to helping many populations survive and perpetuate.

As human development shrinks habitats into isolated patches, wild populations decline. Cultural attributes, such as birds' songs, simplify. In a scientific article titled "Erosion of Animal Cultures in Fragmented Landscapes," researchers reported on a songbird that lives in North Africa and Spain, Dupont's lark. The researchers note that for populations of these larks, "isolation is associated with impoverishment" of their vocal vocabulary. In isolated populations, "song repertoires pass through a cultural bottleneck and significantly decline in variety." Unfortunately, isolated larks are not an isolated case. Researchers studying South America's orange-billed sparrow found that the bird's "song complexity" — number of syllables per song and song length — deteriorated as humans continued whittling their forests into smaller fragments. Might these impoverished creatures *feel* the richness of their life draining to shades of gray? I hope not. What I do hope is that conservationists can advance the case for preserving wide cultural diversity and ease the public out of a perilous satisfaction with precariously minimal populations. Cultural elements erode as habitats shrink, but humans are poor at valuing diversity. Species recovery goals are sometimes too small to save a species, not to mention what the species has learned culturally about how to survive.

I'm not just talking about a few songs. Survival of numerous species depends on cultural adaptation. How many? Likely very many. We're just beginning to ask such questions. But the preliminary answers indicate surprising and widespread ways that animals survive by cultural learning.

And what of the beautiful macaws who have expanded my soul? Beauty itself, of course, is a realm of diversity. If cultures are put under pressures that erode and simplify them, what are the implications for the continued evolution of beauty — or for the survival of the beautiful? As Shakespeare wrote, "How with this rage shall beauty hold a plea?"

BEAUTY:
TWO

Gaby first learned about parrot personality when, as a teenager in Lima, Peru, she acquired two blue-headed *Pionus* parrots. "Malu was friendly — 'Scratch me' — but not interested in learning," she recalls. "Luis was avid for new tricks, always watching your hand, wanting to learn. Luis was *smart*." He had calls meaning "Predator overhead" (applied to raptors but also to plastic bags in the wind) and "Predator on the ground" (most frequently a neighbor's cat).

Gaby is telling me this as we're walking through the rainforest to one of several nests she's been monitoring. She's wearing tall rubber boots, camo pants; a kerchief confines her black ponytail.

Hearing about her pets is interesting. But the *forest* —

The forest overwhelms. A wall of green in all directions, from the ferns of the ground to the leafy ceiling of the high canopy, every level occupied by growing living things, life-forms all scrambling, struggling, competing. Trees with straight dark trunks, spiny trunks, blotchy trunks. Some variegated, some moss-covered, many vine-encumbered. Giant fig trees splay head-high finlike buttresses that emanate from the trunk like ribbony garden walls. Others drop so-called prop roots downward from the trunk like thick support-ing cables. Strangler figs climb the trunks of their victims in their initial vinelike sprint for sunlight. *Capironas,* or "naked trees," shirk off the stranglers by shedding their bark, a continuous slow-motion scuffle with their at-tacker. A multitude of saplings bide their time in shade until a giant falls and sunlight trig-gers their one big chance to surge. Trillions of leaves: pointy, round, flat, ridged, fluted —

Some palms are spare, others lush. The ground is littered with fruits and nuts and seedpods, and riven with fungi enough to pulverize it all as seasons pile into deep ages. The inanimate produces the animate as vegetable grows the animal, grows even macaws, who know their world so intimately. Everywhere bugs and spiders endeavor to pry their living from the world, and the forest floor is heavily patrolled by ants of many kinds and sizes, whose food and work all dif-fer from one another's, the various species as specialized as trade unions in a vast city. Every few minutes some nearby trees get rustled by spider monkeys, howlers, titis, tamarins, capuchins, or squirrel monkeys. It is magic enough even before the morpho butterflies, whose wings resemble dried leaves when folded in prayerful rest, rouse themselves, providing another shock of beauty as they weave their electric blue flickers through sun shafts and shadows. This bewildering green blizzard, this remaining piece of the original world, must be the most confusing profusion of vegetative power on Earth, a great megalopolis of lives and of Life.

"I'm sorry," I offer. "What were you saying?"

"Just, my pet parrots when I was a kid —"

"Okay; I heard that part."

"That's all."

We continue quietly, Gaby and I, moving for minutes through the wholly wondrous.

Our destination is a gigantic *Dipteryx* tree. Its massive trunk has hoisted its canopy far above the surrounding forest. Such standouts are called "emergent" trees. Those in the genus *Dipteryx* can live a thousand years. Natural nesting cavities in such trees might last decades, perhaps centuries, producing hundreds of scarlet and red-and-green macaws — feathered fruits of the tree itself. Macaws like to nest in huge specimens like the one we're heading for because they offer broad, open views. The birds don't like trees full of vines or low branches that can hide climbing predators.

But humans have heavily targeted *Dipteryx* species for flooring wood and for charcoal. So throughout vast areas, the great, gnarled, multi-centenarian giants riddled with cavities needed by birds and other animals have gone crashing to the forest floor, taking macaw populations down with them.

When we get to the tree, its silent immensity and motionless power induce awe. Way up in this tree, a good 120 feet or so, sits a large wooden box, about four feet high by two feet wide and deep. A giant birdhouse for macaws. Artificial boxes are a partial solution to the housing shortage. Only scarlet macaws seem easily amenable to these houses, however. Red-and-greens strongly prefer natural cavities.

This was Tabasco's nest for nine years. He and Señora Tabasco lost it to a younger pair, the current residents, in a violent fight four years ago. That means, of course, that besides love, there is violence among macaws. The new pair then lost it for a year to another pair, then pried it back again, showing that the pair bond lasts even when there is no breeding.

When we arrive, the female is in the nest, and the male is off somewhere. Yet there are two macaws in the crown of the same tree. They are billing, preening each other, and fooling around. Gaby says, "There is no way that the female in the nest doesn't know who these two are. She would not be sleeping." They could be last year's young from this nest, Gaby hypothesizes. She scrutinizes them through her binoculars and adds, "They have really young faces. The skin is smooth — it's not wrinkly as in old macaws. They are being very vocal. And the mama is sleeping. She'd be looking out of the nest — at the very least — if she didn't know them." So, Gaby thinks they're last year's chicks, visiting.

Macaws don't defend food. Many forage in the same trees when fruit is abundant. But macaws do defend — vigorously, violently — nesting hollows. They must.

Seldom does a pair find an empty, undefended, and *desirable* nest. That is partly because nest sites are few, and partly because birds want proven nests, not just empty holes. Not every hole is equal. A vacant hole might have a good reason to be vacant. Some cavities that seem quite suitable for nesting are accessible to predators. Successful predators may return, and at such a site chicks and eggs will repeatedly disappear. Other cavities might look good but turn out to occasionally fill with rainwater.

Some "bad" nests are actually good. One natural nest hollow, a big, deep tree cavity dubbed Vaginito for its shape, has bats roosting on its ceiling. The bats poop on the macaws, and the whole nest seethes with roaches. Yet it succeeds, and the birds fight over it. The best way for macaws to know if a hole is good is that it's been successful in the most recent season. This means that most pairs acquire their nest through a hostile take-over.

A good nest, one that fledges chicks every year, gets challenged at least once a season. Takeover attempts can happen at any time in the nesting season, but more fights occur just as parents are hatching chicks.

Challenges are not always violent. The residents' threat calls can be enough for all parties to size each other up. The birds understand each other's goals: that the challengers are trying to take the site and the nesters are trying to hold on to it. Sometimes, a mere threat and chase routs challengers. If intruders recognize strength in the nest holders, there is no physical contact, no real fighting.

The thing is, they are making crucial assessments of each other. Intruders may detect youth and inexperience or old age. It often appears that they are thinking, making judgments.

If a challenging pair detects a weakness, they may think they have a good chance. And if the nesters have no intention of going quietly, then — the fighting is brutally violent.

Some birds go right for the face of rivals, and a well-landed bite can be devastating. It appears that there is, at least sometimes, intent to kill. A breeder and an intruder may find themselves together inside the nest, fighting viciously in confined quarters. The screaming is intense.

Fights usually last hours, often a day. Fights are bad for chicks; they do not get fed for the duration of the battle. A small chick that does not get fed for an entire day will die. Researchers here have seen breeders at the entrance of the nest for two or three days fending off repeated takeover attempts. During those sieges, the parents never fed their chicks. And though the chicks in those cases were as old as two months, they starved.

When an attacking pair takes over a nest with chicks, they never adopt them. Their takeover anticipates the *next* breeding season. It's all about, and *only* about, the overtaking pair.

When Avecita and her mate succeeded in seizing a nest from a resident pair, it was late in the breeding cycle. The deposed pair had two chicks about fifty-five days old, large and well feathered.

"Avecita destroyed them," Gaby says. "Quick. She went inside and came out with blood all over her beak."

Avecita and her mate — he had a particular black mark on his beak — firmly held the nest for seven years. Lots of macaws are energetic nest defenders. But Avecita was,

Gaby says, "particularly physical. Very aggressive when we climbed, and a successful fighter."

Then she lost a fight, an eye, and her nest.

Several members of Gaby and Don's team are already assembled here, beneath the nest in the huge *Dipteryx*. They have arranged ropes, and one man of the team is donning climbing gear. The researchers make a daily round of such nests to check who has hatched and to chart chick growth and survival. Obtaining such basic information on macaws entails slow, exhaustive work. I used to study ground-nesting seabirds, and to get similar information, I'd look down into a nest, jot down the relevant information, then take a couple of steps to the next nest; I could monitor hundreds of nests a day. Easy.

This is difficult. As the climber ascends, the agitated adult macaws begin growling. One bites the wooden box, demonstrating what that beak can do. Both parents are present, and very annoyed.

"Last year," Gaby notes by way of contrast, "we had macaw researchers visiting from Mexico. The poaching by people for the pet trade there is so bad that in some years, every chick in every nest they monitor gets stolen. So in Mexico, when people show up, the macaws leave. The Mexicans were so surprised at how our macaws stayed, defensive and defiant. I told them, 'Here, we meet as equals.' "

Scarlet macaws live from southern Mexico through Amazonia, one of the largest ranges of any parrot. Being so widespread, they're in no imminent danger of complete extinction. But in many areas, scarlets have suffered drastic declines and regional disappearances. Throughout Central America they're endangered.

Worldwide, nearly a third of parrots are already reduced to low or rapidly declining populations. Several have entirely disappeared in recent times. In 2000, Spix's macaw was declared extinct in the wild. Farming, logging, the cage-bird trade, killing for the cook pot, killing because some farmers consider parrots crop "pests" — all of these factors add up to widespread troubles for parrots.

The climber is just reaching the nest. It's hot and sticky and buggy down here, but he's feeling the winds aloft and enjoying a panoramic view of rainforest unbroken to the far horizons.

The parents begin a crescendo of protective screaming that builds to high intensity.

The climber has opened the door to the nest and is reaching in. He's met by more than the rich macaw smell. One of the adults is there, focusing all attention on lunging at his hand and face.

Gaby adds that the Chicos, having been hand-raised, had no inhibitions about physically attacking researchers at their nests. "It was scary," Gaby recalls. "It's a big thing jumping on you with really mean intentions." Occasionally a Chico would simply latch onto a researcher, sinking that giant bill — which can easily crack a Brazil nut — into, for instance, a shoulder. "Don can tell you how awful the pain was," Gaby adds. "The females were *really* bad. Especially Chuchuy; at the house you can give her some yogurt and she licks from the cup and you think you are friends. But then you climb up to her nest and she tries to cut the rope you're climbing."

"And what would she say about you?" I counter, just to tease. "She comes and eats with you and thinks you're her friend — and then you invade her nest, barging in and removing her chicks and handling and weighing them every day. Is that polite? What's a mother to do?"

High above our heads, the researcher puts the large, twenty-five-day-old chick in a bucket and lowers it gently to the ground. He'll wait up there while the team weighs and measures and assesses health. The adults will stay a few feet away, threatening and screaming occasionally. They know the drill; they've been going through this daily since their first egg hatched.

259

∎ ∎ ∎

The saying "Know your enemy" goes back to Sun Tzu's *The Art of War*, around twenty-five hundred years ago. In a famous series of studies, John Marzluff and his team caught crows so they could tag them with leg bands. After being marked, measured, and released, the crows recognized those people whenever they walked across the university campus, and they'd loudly scold them. Marzluff did not want to upset the crows every time he and his students strolled the campus; nor did he wish to have indignant crows dive-bombing them whenever they were afoot. So to avoid being yelled at for years afterward, Marzluff and his helpers took to wearing masks during captures.

Nine years later, researchers donned the masks of the nasty, dangerous crow-catcher beings. Crows still reacted to the masks, including crows too young to have been around during the bad old days of captures. Those with memories of the terror, who'd had the image of their persecutors seared into their mind, demonstrated to naïve individuals that these were dangerous folk. The naïve ones, having never had a bad experience themselves, learned from elders to harass the masked researchers. As the song says of human children, "You've got to be taught to hate and fear." And so with crows.

Crows remember good things, too. In fact a mysterious, almost mystical thing about crows is evidenced by the many instances of them giving gifts to people who've fed or shown them kindness. Often these gifts are shiny or colorful human-made objects. A culture of fear and a culture of — generosity? Do they *intend* to reciprocate? Hard to say with certainty, but it seems so. This might not be culture like humans have. It's culture like crows have.

The angry macaws trying to protect their nest constitute a hazard aloft for Gaby's climbing assistant, but the ground we're on isn't entirely safe. We watch streams of hunting leaf-cutter and army ants. Leaf-cutters will eat your pack if you leave it. Army ants will eat you. Bullet ants have a bite like slamming a door on your finger. Gliding ants who find your pack unattended will swarm it for your sweat. Sweat bees like to crowd your ears and nose. They don't sting. They bite. Once, while a climber was in a tree at a nest, Gaby grabbed a rope — and immediately got a life-threatening viper bite. The antivenin was at the lodge. She knew she was not "supposed to" run for help; you don't want a racing heart to circulate the venom at speed. "What choice did I have?" she asks. Her team met her halfway. She's here telling the tale.

As the bucket reaches the ground, the team veterinarian disinfects her hands to receive the hefty chick. It has several botfly larvae burrowed into its skin. Botflies can be harmless or can kill chicks, depending on their location. A bit of cream over the entrance to the botfly's burrow in living flesh denies the larva its source of air. In a minute, the rice-sized maggot tries to come through the hole for a breath, and a waiting pair of tweezers ends its picnic. No one grieves for botfly larvae.

Today the team has brought a nineteen-day-old chick to be returned *into* its nest with its twenty-five-day-old sibling. The younger chick had been taken into the lab because the parents were not adequately caring for it.

Many nests hatch two chicks. First-hatched chicks almost always survive to flying age. About half of the second chicks die soon after hatching, usually from simple neglect. When I ask why, Gaby says, "If the second hatches around five days later and is much smaller than the first, the parents are more likely to neglect it."

Don and Gaby are learning that if you remove a small, neglected chick, care for it, and put it back about two weeks later, the parents will raise both. Don says, "They might ignore a tiny chick; they're not going to ignore a big, healthy, month-old chick." The technique can be used for boosting population recovery in regions where macaw numbers have plummeted.

At the research center, this small chick received ten days of intensive feeding and veterinary care. Now though, even at nineteen days, the still-naked chick retains the wobbly and unfocused look of an infant.

Chickens and ducklings hatch with a full coat of down, able to walk, follow their mother, begin learning, and feed themselves. At the other end of the spectrum: parrots. Parrots hatch pink, nearly naked, blind, deaf, and helpless. Their closed-up ears and eyes don't open for a couple of weeks, and during that time they look positively *larval*. I've seen many kinds of newly hatched birds: hawks and doves, songbirds and shorebirds, seabirds and others. Some are adorably cute right out of the egg. Parrots — it gives me no joy to say — are the ugliest baby birds I know. It is fit justice that such ugly chicks grow into some of the most beautiful, intelligent, personable, calculating birds in the world. But quality takes time. Scarlet macaw chicks grow more slowly than those of any other parrot species. New-hatched scarlets weigh about an ounce, roughly one thirty-fifth of their adult weight. Adults weigh about a kilo, a bit over two pounds.

The ground team places both chicks into the bucket and sends them homeward, toward the sky. The waiting climber deposits them in their nest, then rappels back to earth.

Will the parents accept or ignore the still-much-smaller second chick?

Gaby is filled with maternal anxiety as we retreat to where a video monitor is set up under a tarp. There's a camera in the nest. Gaby holds a stopwatch and a clipboard and will note all interactions by each adult, as well as which chick gets fed when, and for how long.

She has performed this routine six times previously. All adults have accepted each newly replaced chick after its prolonged absence. In one nest, where a predator had eaten the eggs, Gaby gave the pair a three-week-old chick whose biological parents weren't caring for it. She says, "You should have seen the face of the male, looking into his nest and suddenly there's this big chick in there. He was like, 'What happened here?' " He flew off. The researchers feared that he had abandoned the nest. Half an hour later, he returned bulging with food and fed the young one. Chicks must consistently gain weight to maintain their health. Fifty feedings a day are not unusual. Nesting and chick-rearing is a full-time job, and if a chick is to survive, the attention must be almost constant.

Less than five minutes after we've sent the chicks back into the nest, the male arrives. He inspects the larger chick and then grasps

the smaller chick's bill and gives it a shake —
which looks too vigorous — prompting the
chick to open wide. The mandible of a small
chick is shaped a bit like two hands begging.
It presents a wide feeding target. The male
bends his head and quivers as he pumps food
into the newly replaced chick. This is a great
relief to Gaby.

While the male is feeding the chick, his
mate comes into the nest. Watching, she nods
her head. The nodding could be taken for approval, but actually it's a low-intensity form
of begging from the male: "Looks good; I'd
like some, too."

This feeding lasts three minutes. As a
breeze sways the treetops, Gaby says cheerfully, "It's a rock-a-bye moment."

Soon this female gives the smaller chick a
feeding so prolonged that Gaby, fixated wide-
eyed on the black-and-white monitor, says,
"My goodness. Gonna make it pop."

The larger chick gets fed, too. Then the
mother preens the chicks for minutes on end,
nibbling all surfaces in fine detail. With such
attention, any skin parasites are likely to be
groomed away. She preens the larger chick so
intensively that it tries to get away from her.

All macaw couples are different. At this
nest, the father always feeds the chicks. In
another nest, the father always feeds the
female, never the chicks. One male almost
never enters the nest. At another nest, the

265

male never screams. At yet another nest, the male screams whenever he arrives, and the female comes out to meet him. One female hardly ever comes out of her nest. Some males deliver food hourly, some about every three hours. A male who takes too long might get a scolding. "I was watching once," Gaby says, "and it looked like the female was looking and listening for the male, awaiting his arrival. When the male arrived, the female was *pissed*. She was making all grumpy sounds" — Gaby mimics an annoyed macaw, *Eh-o-eh-o-eh-o* — "like, 'Where *were* you?' "

Young parents often don't incubate their eggs well or care adequately for their chicks. Younger breeding males don't participate much in direct parenting; they feed their mate, and she does the rest. It might take a pair two or three years of nesting to successfully raise a chick to the age at which it can leave the nest and fly. Gaby says, "The young breeders, they make mistakes. Parenting improves by repetition." Experience creates competence. One female, Gaby is telling me, "is a supermom. She can raise anything."

While we continue our vigil under the tarp, each breeze in the treetops rains seeds, flowers, and bugs onto our awning. But the air at ground level does not move. The mosquitoes like it this way.

All the while, other macaws are chattering beyond green veils of forest as, overhead, a hawk turns circles. Gaby explains, "It's not a kind that's a threat to them. So none of them has given an alarm call. But when it's an eagle that's dangerous to them? They all go crazy."

A couple of years ago, researchers arrived at a nest to find the parents screaming just after climbing weasels called tayras had eaten both chicks. "The parents made a sound we've never heard before," Gaby recalled. "It sounded so sad. They spent a few days just mourning like that." Don had told me that one day at the lodge, one of the Chicos was sitting on a railing. On the ground right beneath him was an orphaned chachalaca — a chicken-like bird — that people at the lodge had been feeding. "All of a sudden, we hear this absolutely earsplitting alarm call. I look up to see a black hawk-eagle coming down straight and fast, probably aimed at the chachalaca but for all practical purposes also flying straight at the macaw." Suddenly, as the Chico on the railing leapt into the air at the sound of the alarm, his wild mate streaked in, no questions asked, and smacked the hawk-eagle. "That was one very unhappy hawk-eagle." Don's take-home: "A macaw will make contact with a dangerous predator that is aiming at their mate. And in this forest, that is about as far as you can go for one you love."

After macaw chicks leave the nest, at about three months of age, they sit around in nearby trees for a week or so. "The young ones are not strong fliers," Gaby explains. "They don't have full-grown wing and tail feathers. And," she emphasizes, "they are *dumb.* They don't understand the world; they don't know what to do." They are clueless about dangers and predators. "They do a lot of sitting and waiting for the parents."

Gaby points out something else: "After the young fledge, the rainy season is over and the amount of fruit in the forest is declining — a lot." The parents continue feeding their chicks for months after they're out and flying. After about a week, the youngsters begin following their parents. Young birds begin learning what there is to eat by watching and sampling what their parents are eating. "They have to learn *how* to be macaws," Gaby says.

In Costa Rica, Sam Williams directs the Macaw Recovery Network. (Most of the Central American forests that macaws need have been felled and burned, largely so that U.S. fast-food burger chains can sell cheap beef.) Williams and his team reintroduce captivity-hatched fledglings to life as free-living birds. Their breeders — mostly scarlet and great green macaws — are rescued captive birds that would not survive release. But preparing young birds to go free in a complex world without parents to follow and learn from, bereft of that cultural guidance, is a slow, risky gambit.

In all free-living parrots that have been studied, nestlings develop individually unique calls, learned from their parents. Researchers have described this as "an intriguing parallel with human parents naming infants." Indeed, these vocal identities help individuals distinguish neighbors, mates, sexes, and individuals — the same functions that human names serve.

Sam tells me that when he studied Amazon parrots, he could hear differences between them saying, essentially, "Let's go," "I'm here, where are you?," and "Hello, darling, I just brought breakfast." Researchers who develop really good ears for parrot vocalization and use technology to study recordings have shown that parrot noise is more organized and meaningful than it sounds to beginners like me. In a study of budgerigars, for instance, birds who were unfamiliar with each other were placed together. In groups of unfamiliar females, it took a few weeks for their calls to converge and sound similar. Males copied the calls of females. Black-capped chickadee flock members' calls converge, so that they can distinguish members of their own flock from those of other flocks. The fact that this happens, and that it takes weeks, suggests that free-living groups must normally be stable, that groups have their own identity, and that the members group-identify.

Group identity, as we see repeatedly, is not exclusively human. We've seen how sperm whales learn and announce their group identity. Parrots and even young fruit bats learn the dialects of the crowds they're in. Ravens know who's in, who's out. Too many animals to list understand what group, troop, family, or pack they belong with. In Brazil, some dolphins drive fish toward fishermen's nets for a share of the catch. Other dolphins don't. The ones who do sound different from the ones who don't. And orca whales, the most socially cultured non-humans, have layered societies of pods, clans, and communities, with community members knowing the members of all their constituent pods but each community scrupulously avoiding contact with members of another community. Group identity, often considered a hallmark of human culture, turns out to be widespread.

Many young birds need to learn much by observing their parents and elders, and parrots probably need to learn more than most. That's why trying to restore parrot populations by captive breeding and reintroduction is tricky and fraught. It's not as easy as training young or orphaned creatures to recognize what is food while they're in the safety of a cage, then simply opening the door. "In a cage," Sam Williams says, "you can't train them to know where, when, and how to *find* that food, or about trees with good nest sites." And a landscape is complex and ever changing. "Just throwing birds out when we haven't prepared them for survival would be unethical," Williams believes. Worse, it might not work. The prospects for survival of released individuals are most severely undermined when there are no free-living elder role models. A generational break in cultural traditions hampered attempts to reintroduce thick-billed parrots to parts of the southwestern United States where they'd been wiped out. Conservation workers could not teach the captive-raised parrots to search for and find their traditional wild foods. If they'd had parents or a social flock, they would have learned these things as a matter of course.

Elders also appear important for social learning of migratory routes. Various storks, vultures, eagles, and hawks depend on following the cues of their elders to locate strategic migration flyways or important stopover sites. These could be called their migration cultures. Famously, conservationists have raised young cranes, geese, and swans to follow microlight aircraft, which then acted as a surrogate parent on their first migrations. The young birds culturally absorbed knowledge of the routes and used it in later seasons on their own self-guided

271

migrations. Four thousand species of birds migrate, and Andrew Whiten of the University of St. Andrews in Scotland speculates that following experienced birds could be "a potentially very significant realm of cultural transmission."

Young mammals, too — moose, bison, deer, antelope, wild sheep, ibex, and many others — learn crucial migration routes and destinations from elder keepers of traditional knowledge. Conservationists have recently reintroduced large mammals in a few areas where they've been wiped out, but because animals released into unfamiliar landscapes don't know where food is, where dangers lurk, or where to go when the seasons change, many of these translocations have failed. Among fishes, guppies, bluehead wrasses, and French grunts introduced by researchers into a preexisting group followed residents to feeding and resting areas. The newcomers continued to use these traditional routes after all the original fish from whom they had learned them were gone. We humans inherit ways to dress, and foods to eat, and the music we enjoy. Often we are not even taught these things. From birth, we are simply immersed in our elders' ways.

When you simply look at free-living birds and other animals, you don't usually *see* culture. Culture makes itself felt when it gets disrupted. Then we notice that the road back to reestablishing cultures — the answers to the questions of "how we live in this place" — is difficult, often fatal. In Williams's operation, former pets have proven to be the worst candidates for release; they don't interact appropriately with other macaws, and they tend to hang around near humans.

Williams describes his procedure as "very much a slow release." First, his team trains the birds slated for rewilding to use a feeder. With that safety net, the birds can explore the forest, gain local knowledge, begin dispersing, and try eating wild foods. Some rescue programs declare success if a released animal survives for one year. "A year is meaningless," Williams says, "for a bird like a macaw that doesn't mature until it's eight years old."

I ask what they're doing for those eight long years.

"Social learning," he immediately replies. "Working out who's who, how to interact, like kids in school."

To gain access to the future, to mate and to raise young, the birds Williams is releasing must enter into the culture of their kind. But from whom will they learn, if no one is out there? At the very least, they must be socially oriented to one another. To assess the social abilities of thirteen scarlet macaws who were scheduled for release, Williams and his crew documented certain behaviors, including how much time each bird spent close to another bird and how often they initiated aggression. When the bird scoring lowest for social skills was released, he flew out the door and was never seen again. The next-to-lowest candidate didn't adapt to the free-living life and had to be retrieved. The third-lowest social scorer remained at liberty but spent a lot of time alone. The rest did well.

All of the above sums to this: a species isn't just one big jar of jelly beans of the same color. It's different smaller jars with differing hues in different places. From region to region, genetics can vary. And cultural traditions can differ. Different populations might use different tools, different migration routes, different ways of calling and being understood. All populations have their answers to the question of how to live.

Bottom line, said Williams: there is much going on in the social and cultural lives of species such as macaws that they understand — and that we don't. We have a lot of questions. The answers must lurk, somewhere, in macaw minds.

"Sometimes a group will be foraging in a tree," Williams adds. "A pair will fly overhead on a straight path. Someone will make a contact call, and the flying birds will loop around and land with the callers. They seem to have their friends."

As the land, the weather, and the climate changes, some of these differences will turn out to be the answers of the future. Others will die out. If diversity remains in these cultural pools — and there is sometimes more diversity in cultures than in gene pools — the species' survival will become more likely. If pressures cause regional populations to blink out, the species' overall odds of persisting will dim. With populations blinking out, the odds of maintaining the rich tapestry and beautiful panorama of Life on Earth become a more tenuous proposition, increasingly diminished.

Williams's goal is to reestablish macaws where they range no longer. It often takes a couple of generations before the descendants of human immigrants learn how to function effectively in their new culture; it may take two or three generations before an introduced population of macaws succeeds at becoming wild. In other words, macaws are born to be wild. But *becoming* wild requires an education.

For fifty million years, parrots flew where the mood took them. No parrot suffered the indignity of clipped wings, went insane with loneliness, plucked himself bare, or ached for affection and normal society. No one talked gibberish to parrots or taught them foul language. For fifty million years, no parrot heard a chain saw, or fled a nesting tree being cut down, or saw its forest felled and burned and filled with scrawny cattle.

Parrots were made for a world that made them and provided everything they needed. The world knew who parrots were. Parrots knew their world. The world of parrots was among the richest and the most *beautiful* realms of that original world.

BEAUTY:
THREE

It is not becoming to humanity that I remain silent while birds chant praises.
— Saadi, 1265

Night still knows how to be night here in the rainforest. The swollen moon reflects ample light, but the forest path is a hazard of roots, so Don and I switch on our headlamps and tramp through the shadows of great trees to the river's edge. As the trills of pygmy owls fall away, howler monkeys begin revving up their predawn prayers. From deep within their big throats, sounding like a cross between Himalayan throat singers and gravel in a blender, they send through the treetops sustained roars and reassertions of their claims, reconvening the world.

When we step into the wooden boat, the lingering darkness seems to amplify the sound of our footsteps. I watch the Southern Cross through large-leaved *Cecropia* trees as we get into the swift flow of the Tambopata River and motor a short distance upstream.

The sweep of my light reveals that the air just above the water is flecked with bats, and it makes small fish jump along the black water.

Beaching the bow on a small island, we step onto the silty bank and walk to the island's far side. The shifting river abandoned that farther channel about twenty years ago. In its place has grown a reedy marsh. Across this marsh are bluffs of clay, eroded and exposed in the rushes and floods of the river's former presence here. Atop the bluff the thick rainforest erects a sudden vegetative wall. Rising beyond even this, a giant of a kapok tree pierces the forest skyline. Sometimes called a ceiba, sometimes considered sacred, this tree overspreads the jungle below, its umbrella dwarfing even the other rainforest monarchs, an emperor among royalty, a great and generous presence. Over the towering kapok tree shines one bright planet, the indelible sight of Venus above the Amazon.

We get situated as the night goes to sleep. We're seated, comfortably enough, with our binoculars and a good view across the reedy old river channel to the russet and beige layers and crumbling folds of the clay bluff. We ready our notebooks and sip some coffee, waiting for day to wake.

As the night's shades rise and the eyelids of morning begin to open, the world retells its creation story. I was expecting something like the explosive dawn chorus of springtime in the Northeast. But here the sun comes up slower. And so does the chorus. It starts with chanting.

Dawn's main meditation soundtrack begins with razor-billed curassows, who call like this: Inhale. Hold. Exhale, and with each exhalation send out a deep, resonating *Whoooohh* like a superlatively large dove. As the curassows coo their oms into the ravine, their meditative tolls get overlaid by three soft, insistently plaintive whistles, the timorous affirmations of an undulated tinamou. These two calls somehow reach each other inside my mind, meshing into one of the most beautiful things I've ever heard. The slow rhythm is like the whole forest breathing.

Don, who has brought me into this new magic, notes the approach in the predawn gloam of two scarlet macaws who are just silhouettes screeching. Two additional "scarlets," flying over the tree canopy, remain dusky shapes against thin clouds.

While the sky is still blue-gray, degrees of illumination cue others to appear on the soundstage. Oropendolas begin calling, their notes sounding like big drops of thick liquid. Motmots begin adding their rhythmic tom-tom beats. Their notes are so meditative and soothing that they set my mind adrift, like a boat in long swells. Rise. Subside. I am afloat on an emerald sea amid shoals of birds. The volume comes up, comes up, until we are wrapped in a shimmering surround of songs and calls, some melodic, some emphatic.

It literally dawns on me that countless generations of singers who've continually come and gone have performed this soundtrack of renewed existence daily here for countless thousands of years.

And if all had been as it was supposed to be, there'd be no end in sight.

The expanding light brings waves of added chatter. Titi monkeys who sound birdlike, the occasional flute of a white-throated toucan, the whistle of a hawk. And much else. A great antshrike chitters and squeaks and rattles so full of energy that I can't decide whether he seems filled with jubilation, trepidation, agitation, or anticipation. Perhaps he knows no such differences. A kiskadee — white eyebrows and yellow belly — pops up in a stand of bamboo. He regards us not at all, disregards us entirely. This is very reassuring.

Don mentions that we are in the sweet spot here. First of all, the forest remains intact. Second, in places where wildlife is hunted, it's either already been shot out or the survivors hide; you don't see them. Where there are no people at all, wild animals are wary; you don't see much there, either. But where people are a consistent presence *and* don't hunt — like here, near the research center and lodge — the animals relax and you get to see them going about their normal lives, doing what they do, being who they are.

As Shakespeare wrote, "I like this place and willingly could waste my time in it." We're near the middle of Peru's Tambopata National Reserve and Bahuaja-Sonene National Park, totaling 5,275 square miles. Immediately beyond these protected areas, forests are felled and burned by loggers and squatters, and water channels are spoiled and polluted with mercury by illegal gold mining. But for now the reserve and its miracles remain safe-boxed. Tourists' currencies are key to keeping this portion of forest locked from harm.

The question, Don says, "is whether this huge protected area is really big enough to protect the many thousands of parrots that live in it." Is this enough room for viable populations, in the long term?

I want to know everything there is to know about this place. But not right now. For just a few minutes I want to feel what there is to feel, to inhale this new oxygen, notice each sight, sound, scent, to hear each call, distinguish each note in the orchestra, pay attention, and then relax and let the great music wash over me. Only a fraction of such ambitions will be possible. Nonetheless I permit myself the luxury, a few moments' indulgence, while there remains the room and the

281

time for it, in this unruined corner of the world.

The ode to joy that enlivens this planet, our home, generates information, of course. And it generates joy's by-products: mystery and tragedy. The tragedy is that we humans have so overwhelmed ourselves in an "information" age where our own signals crisscross and messages bombard us and everything beeps for our attention that we've stopped listening. The whole world continues calling out.

For a few minutes, I *am* listening. This tropic dawn, slow-handed as love, rushes nothing. It savors its own time as it celebrates its daily rite of passage.

But I have come for parrots.

Just as the streaky clouds get edged with a palette of pinks, a few Amazon parrots arrive on shuddering wings. "Amazon parrot" sounds like one species; there are more than two dozen. These new arrivals are called "mealy Amazons" because their mostly green backs look as if they've been lightly dusted with flour. The first mealies land in favored trees at the far end of the bluff across the marsh. The bluff is about a hundred feet high and takes its name, Colorado, from the clay's reddish hue. There above the clay, the parrots talk talk talk. Meanwhile, a blue-throated piping guan has appeared like a dark, distant dot. Through binoculars we watch the guan chiseling clay fragments out of the ground.

Some animals are social and often travel in pairs even within larger groups. Humans are such animals. So are macaws. Even in flocks, duos keep close company aloft. When one macaw arrives alone, it's unusual enough for Don to comment, "That scarlet macaw probably has a mate back home in a nest."

I'm having a hard time understanding how Don is differentiating the various macaw species, and even the smaller parrots, at such distances in low light. Seeing me struggling to spot differences even in the big macaws, Don says, "Scarlet macaws have longer tails that are a bit wiggly in flight." He points to four birds flying below the tree line above the clay bluff. "The red-and-green macaws have stiffer tails, and their heads are proportionately larger."

Okay. Yes, I can see now that the biggest macaws' silhouettes are often quite enough for a positive ID. But all these smaller parrots? No; to me they look much too similar.

The sun has not yet attained sufficient elevation to send direct light shafts onto the forest. But it is now striking the highest fliers, suddenly making the scarlet macaws coming from high above the forest canopy look like lit embers crossing pale space.

It's a brightening morning with a blueing sky.

"Perfect clay-lick weather," Don says cheerfully. *All* the parrots are commuting in for one reason: to eat clay. No other "clay lick" in the Amazon attracts so diverse a complement of parrots; seventeen species come here, including six types of macaws: the big red-and-green macaws, scarlet macaws, blue-and-yellow macaws, and three smaller species. Chestnut-fronted macaws are no bigger than Amazon parrots. It's not size that earns a bird the enigmatic label "macaw." Faces featuring a lot of bare skin qualify one for the designation.

As commuters continue arriving, their sounds build tidally, a slow crescendo. For millions of years, the main sound in Earth's atmosphere was the aural flood of messages birds exchanged. Birds remain the most noticeable free-living animals, coloring the world with dazzling palettes of sound. Go almost anywhere. Subtract the motors. Listen.

As more and more parrots come filtering in at an accelerating pace, we are hearing far more than I am seeing. And now I am having a problem. A few minutes ago, I was learning the calls of birds new to me: curassows, tinamous, guans; the kiskadees and jays. I hear them all, and they all sound different. Each sings a real song, easy enough to learn and to distinguish from all other callers. But in the various parrots and parakeets — probably a dozen species here already this morning — and in the macaws I've come here to focus on, I perceive only noise building to one indistinguishable din. They seemingly have no repeated call or song, as so many other birds do. Shakespeare's Juliet mistook a lark for a nightingale. Or so she insisted; it's hard to mistake those birds. But to me all the parrots simply sound similar, and harsh. To me, they all squawk. I can't make *any* sense of parrot noise. I find this disconcerting.

Don expresses measured sympathy. "It's hard," he acknowledges, adding, "scarlets have more of a dog-growly voice."

Lifting his pencil from his clipboard, he points for my benefit. "Hear the red-and-greens?" he says in all earnestness. "They have the most distinctive call of all the macaws here."

If you imagine being a fast-moving parrot in a low-visibility rainforest, you can appreciate that it's crucial to hear and recognize individuals, and to be heard as an individual. Experiments with recordings show that females distinguish calls of their own mates and will come out of a dark nest cavity for the sound of the right male. In fact, their individual voice differences can be seen when recorded sounds are turned into graphs called sonograms. When parrots are being so noisy as they move, a lot of what they're doing is staying in touch and knowing who's where.

But as a beginner here, I can't yet sort through all their sounds.

The smaller parrots, especially the little cobalt-winged, arrive in flocks so cheerfully disorganized that their own scatter identifies them. For people who've seen only caged or wing-clipped birds, it may seem surprising that free-living parrots are *highly* competent fliers. They fly right, left; they land, then switch from one tree to another. A small flock of orange-cheeked parrots flashes by, sounding full of their own lives.

In this bewildering open-air cathedral, my struggle to see and hear so many distinctions is frustrating. But it is also thrilling. The parrots' excitement in being alive is contagious; it inoculates the very air.

"Dusky-headed parakeet arrival." With each new shot of birds, Don jots down their time and numbers. "White-bellied parrot arrival." He tells me that white-bellieds nest communally; they travel in stable family groups of six or eight. "Blue-headed *Pionus* parrots arriving straight across." With due diligence, Don makes his notes. Without looking up he says, "Another small group of white-bellieds going left, low." Don hears their identity.

If Don's ability to detect the different macaws by voice is impressive, his ability to pick out the different smaller parrots' calls is so far above my pay grade as to seem preternatural. He is so skilled it strains credulity; *how* is he doing it?

Busy taking notes, he says only, "With captive birds, they're all over the place, as far as the sounds they'll make. Wild ones, you can hear the different voices."

Actually, I can't. But the fact that he can means the birds must find it breezily easy to hear all the differences. How much information, and emotion, and intention might these hot-blooded flocks be sharing?

Parrots' brains have a vocal system within a vocal system. Think of it as a "core" and a "shell." The parrot "core" system resembles the song systems of songbirds and hummingbirds. The "shell" system is something only parrots have. Among various parrots, relative sizes of the core and shell systems differ greatly. Species that are better at mimicking have noticeably larger shells relative to core regions.

Parrots, "songbirds" (roughly half of the world's birds, about four thousand species), and hummingbirds can learn their vocalizations and can vary them based on individuality and what they hear from others. The vocal repertoire of other birds such as gulls, hawks, owls, herons, loons, and albatrosses appears to be more preprogrammed, as human smiling is. The ability to learn new and different vocalizations exists — as far as is known — across an odd smattering of other animals: humans, of course, but also bats, elephants, dolphins and whales, seals and sea lions, goats, damselfish, Atlantic cod, perhaps certain insects. With a list like that, there must be others. In some species, new calls arise and others fall out of use. Or some components change while some remain, as when humpback whales collectively change parts of their song while continuing to sing other parts.

The vocalizations often serving dual purposes of territorial claim and competitive advertising for a mate are usually called "songs." "Calls" are for staying in touch, close contact or bonding, identification as when parents and young are searching for each other, and so on. The terminology can be a little confusing. A loon isn't a "songbird," but its famous call is functionally a song. Owls sing with trills or coded hoots. A crow *is* a "songbird." It's non-musical caws are its song, but those caws also serve other social purposes.

Parrots evolved their vocal learning systems very early in their history, roughly thirty million years ago. Though the earliest humans heard plenty of birdcalls, people didn't start thinking scientifically about birdsong until the 1800s. When they did, they got off to a bumpy start.

Charles Darwin suspected that birds pass songs along from generation to generation, as young birds learn from elders. He was presciently observant and intuitive. But Darwin's style — richly collected anecdotes, big-picture insight, and putting all animals on the same playing field by applying the same descriptive language to all creatures — fell from scientific fashion. By the 1920s, observing nature seemed old-fashioned and sloppy. Many scientists moved animals into labs and cages for controlled experimentation. Darwin had no problem saying that animals had fun playing. But suddenly only people could "play"; a behaviorist might say that non-humans could merely "engage in affiliative activity."

Carefully detailed experiments both expanded and constrained our understanding about the nature of nature. Information was gained. Though as Henry Beston wrote, the price was seeing "a feather magnified and the whole image in distortion."

Experimentation gave us useful concepts. But concepts too often got hardened into dogma.

Dogma isn't scientific; still, scientists are only human and they, too, can be rigid. Many scientists came to believe that all behavior other than humans' was instinctual, none learned. Dogma about behavior began substituting for observation. "A mockingbird does not mimic songs but only possesses an unusually large series of melodies which it calls forth," asserted one J. Paul Visscher in 1928. He concluded that "these birds inherit various neural patterns." Actually, the slightest study of mockingbirds informs us that they learn and mimic the songs and sounds around them. It took only one individual mockingbird to falsify Visscher's claim. In the 1930s, Amelia Laskey plucked a nestling from its parents and raised it in her home. Honey Child learned to mimic the birds outside her home, the whistle of Laskey's husband summoning their dog, the washing machine, and the mailman calling.

Songbirds learn their songs, and parrots learn calls. Their vocalizations are not innate. But — and this is a big but — birds, like humans, do possess a genetic template that limits, and perhaps dictates, what they *can* learn. There is latitude — room for variation and differences in song richness and regional dialects — yet not complete freedom. No crow will ever sing like a canary. Instead, crows can make an array of sounds and even do some mimicking. At the far end of the spectrum, lyrebirds take copying to extremes, mimicking chain saws and car alarms. Parrots use their sound-copying ability to identify themselves and others.

A human baby cries instinctively, without

290

learning. Many nestling birds utter instinctive food-begging calls. Just as baby humans *learn* to speak the human language they hear — but cannot learn to sing like a lark — a lark listens to the song it hears most: its father's.

Very early on, the nestling's brain begins creating cells in regions specialized for fine discrimination of sound. As in human language acquisition, young birds have a window of time — a couple of months, in their case — when learning comes easily. In some parrots and a few other species, this window stays open longer and closes more slowly. Some birds can learn their species' repertoire only when young. (Isolated experimentally, a young bird will usually utter a stunted, stilted version of its species' normal adult song and social vocalizations.) A nestling will begin babbling snippets of adult notes it is learning by hearing. This is quite analogous to a human baby's babbling. Meanwhile, researchers working with finches have recently revealed that adults "alter the structure of their vocalizations when interacting with juveniles in ways that resemble how humans alter their speech when interacting with infants."

Those fine-tuned social interactions alter brain activity, helping focus attention on song, speeding vocal learning. Eventually the song (or songs) and calls that emerge will be the ones the bird has a mental template for learning and the throat for singing. It's like reading music and playing it. With practice, the bird develops control.

Various musicians will play and interpret the same piece differently. And where learning is involved in birds' vocalizations, different versions crop up. Regionally different vocalizations are sometimes called "song traditions," but the better and more commonly used word is "dialects."

More than a hundred studies on dialects in birds have been published; it's definitely a thing. And it's not just birds who create regional variations but a wide array of animals. As the great conservation biologist Tom Lovejoy commented to me, "Any social animal creates their own dialect, whether it's a flock of parrots or kids at summer camp."

Including some fish. "Cod particularly," said Steve Simpson of the University of Exeter, "have very elaborate calls compared with many fish." You can easily hear differences in recorded calls of American and European cod. "This species is highly vocal, with traditional breeding grounds established over hundreds or even thousands of years, so the potential for regionalism is there."

Yellow-naped Amazon parrots in Costa Rica use three regional dialects (north, south, and Nicaraguan). They learn the dialect of the locale where they are raised. The birds are all from the same gene pool; the different dialects are cultural. When individuals move into other dialect areas, they lose their natal tongue and — when in Rome — acquire the local parrot parlance. And things change; dialect boundaries shift over the scale of around ten years. That's not much time in the multi-decade life span of these parrots. The talent some parrots show for learning human words reflects their natural ability to learn and distinguish changing parrot dialects during normal life in the wild.

Simply put, traditions vary from place to place and time to time. We all know that human speech and music vary and change. And it turns out that birds' songs often change, too. When a scientist played twenty-four-year-old recordings of singing male white-crowned sparrows at the same location where she'd recorded them, they elicited half the responses they had at the time they were recorded. The birds' responses show that changes in the dialect lead to changes in listener preference, a bit analogous to humans' changing tastes in pop music. And as with humans, preferences can affect whether a particular bird will be accepted as a mate (white-crowned sparrows singing a local dialect father more offspring than do singers of unfamiliar dialects, indicating that females prefer a familiar tune). When other scientists studied indigo buntings' vocalizations, they found that variations in the songs allowed researchers to identify who a given indigo singer had learned from. That's a bit like hearing the style of a teacher in the musical performance of their student. On the other hand, errors and innovations are what move the dialect culture along. Genetics and learning, nature and nurture, work together.

Bird researcher Tim Birkhead says that the study of how birds acquire their song "has provided the most compelling evidence that there is no nature-nurture divide: genes and learning are intimately interconnected in both birds and human babies." It was through the study of the brains of singing birds, he notes, "that we began to realize the huge potential for the human brain to reorganize itself and form new connections."

Earning control over sounds requires practice for babies, for musicians, and for birds, too. The incentive for practicing over and again is that, as with a child or a musician getting it right, birds' brains give them satisfying doses of dopamine and natural opioids when they do. The neurotransmitter dopamine is involved in both human speech and birdsong. Dopamine is both an enabler and a motivator. In songbirds, singing causes the brain to release dopamine. People who lose the neurons that make dopamine are referred to as having Parkinson's disease, and Parkinsonian patients who receive high-dose dopamine-replacement therapy occasionally develop compulsive humming and singing.

Dopamine is released into a bird's brain during singing (especially if a male bird is singing directly to a female he is courting). And because dopamine is a kind of "feel good" chemical, we can answer the age-old question of whether birds sing because they're happy. Simply, birds are happy because they sing, *and* they sing because they are happy.

They don't just sing mindlessly. A bird's mind is entirely involved, in ways physically and chemically similar to our own when we speak or sing. Maybe you knew that; now we know why. When they are singing to warn and to woo, singing feels good. Singing feels good to us too, which is why we do it.

The first scarlet macaws that I can see really well materialize out of mists rising from the forest canopy in the warming morning. Arrayed as two pairs, these four land in the tip of one of the tallest trees atop the bluff, alongside a pair of red-and-green macaws. And when the first sun rays to drill through the mist strike them directly, they all seem to simply ignite. Their coloring is so striking, so exquisite, that it perplexes me. Macaws are so full of exaggeration that they almost taunt us with a question: Why? Why such colors, such voices? Indeed, this calls forth a question so confounding as to be almost mystical: Why do they possess such beauty, and why do we humans have the capacity to perceive it?

BEAUTY:
FOUR

"Look." Don points to a leafy tree. "In that *Cecropia*. Just below the two yellow-crowned parrots. To the right of the two red-bellied macaws there." Three blue-and-yellow macaws just landed. "They've got to be a family. Two parents and last year's chick."

My binoculars find them as the early sun sets their gold bellies aglow. Their foreheads look aqua, and their blue backs grade to indigo on their shoulders, wings, and tail. They have a striking black throat strap, and their white cheeks show elegant black scrawls that differ from individual to individual. The young one, wings quivering in supplication, is practically hugging one adult as it begs. The parent remains impassive.

"Year-olds have a kind of lost look about them," Don says. "They sort of just look around while the adults are very seriously taking in the scene and evaluating things."

It is already getting hot and they are all panting, black tongues moving with their breathing.

Eventually they launch into the air and join a small skein of macaws flying along the clay face to land in favored trees that combine nice safe, leafy cover with the proximity of the adjacent clay that is their destination and objective. But it's not so simple.

"Classic indecision," Don explains. "That lowest macaw is watching four guans on the clay. He really wants to go down but can't work up the guts." Don poses a question: "How difficult is it to decide where to go for dinner? If it's just you and your wife, not difficult — right? But if you're at a conference and twenty people want to go out? No one can decide. Now imagine five hundred parrots trying to decide which end of a clay lick they're going to land on. It's a decision nightmare!" He laughs.

The bluff's russet, gray, and chocolate striations indicate layers varying in consistency and composition. What makes arduously gouging it worth the travel, jostle, and risk: the clay contains sodium. The main draws are two: sodium is crucial for animal health and cellular function, and there is almost no sodium elsewhere in this region.

Don has studied the situation in detail. "They're by a spot where the clay's sodium concentration is six to seven hundred parts per million. Below that is a band with over one thousand, but it's too close to the vegetation, and they don't feel safe enough there."

Sodium is often transported in rain, from ocean-originated mist and vapor. Here, east of the Andes Mountains in Peru's lush Amazon region, weather systems tend to come from the east, and by the time the rain gets this far, it has fallen, evaporated, and fallen again several times, as the water cycles westward across the continent. It arrives here devoid of sodium. The only place in the hemisphere where animals are so sodium-starved that they need to mine it is here in the western Amazon. But here, too, millions of years ago, an ocean laid its salty sediments down for posterity.

And now, in a forest far, far in time and distance from any tidal coast, a Spix's guan, which looks like a black turkey, walks out onto the farthest end of the bare slope, chiseling at the clay. Also on the clay: three blue-throated piping-guans and a couple of chicken-like speckled chachalacas.

None of the parrots care that a guan or a chachalaca considers it safe to be there.

What parrots do, they do in groups. You don't see single parrots on the ground. They wait for reinforcements. In the trees, their numbers build; their noise levels build.

Don says, "You almost get the feeling they're goading each other, trying to encourage someone else to be the first to go down."

Eventually the parrots will judge the presence

of a critical mass. Not yet.

The bigger macaws — slower, showier, louder, and brasher — out-parrot all other parrots. You'd think they'd fear no evil in this vale. You'd think they'd be first. But they are incredibly circumspect. A couple dozen head to one tree, fly to another. They start to go to one spot, but before they land they spiral away in a vortex of blurring color.

Suddenly the Amazon parrots — hundreds — begin fluttering to the ground, covering the bare clay with their green bodies like an instant lawn of birds. Despite shoulder-to-shoulder packing and a few jostles, there are no fights. I can see through my binoculars that the clay where they are is crumbly. The eating is leisurely. Many stand while holding a footful of dry clay, nibbling it like a child with a big cookie.

Meanwhile, the big macaws continue gathering. Their hues and tints set the trees aglow, the spontaneous combustion of their reds, yellows, and blues making them the rainforest's supreme firebirds. Most of two hours passes after the Amazon parrots hit the ground, and still the macaws remain in the branches.

When the scarlet macaws number about sixty, their collective chatter escalates in intensity. They are making a committee-level decision. What's being communicated? Don says simply, "The sound builds as they get up the gumption to go down."

The safe progress of smaller parrots seems an insufficient all-is-clear for the macaws. Likewise, they seem unconvinced by the two violaceous jays who are ambulating unmolested on the clay face. It matters little to them that the guans and chachalacas on the ground are doing just fine. It seems that macaws set higher standards for their own safety and security. It's almost as if they want to be really sure that they won't die for a crumble of clay.

What are they on the lookout for? Crested eagles. Ornate hawk-eagles. Black-and-white hawk-eagles. Forest falcons. A harpy eagle. Cats, including margays and ocelots, jaguars and pumas. Their size does not *fully* exempt them from dangers.

Don says their vigilance pays. There are more than a dozen kinds of birds of prey here, but "in eighteen years, we can count on our fingers the number of kills we've seen." It's nearly impossible for a sharp-clawed hunter to surprise such a sharp-eyed group. "They've developed," he notes, "a very predator-resistant system."

The big macaws remain unswayed by what the smaller parrots are doing. Eventually their group decision to go onto the clay is accomplished by individuals dropping to lower branches, slowly approaching the bare clay slope of the bluff, until a lone blue-and-yellow macaw boldly ventures where no macaw has gone today, onto the clay itself. Once that spell is broken, more land, crowding closely.

Now there are twenty big macaws on the ground. Roughly half are scarlets, half red-and-greens, a few blue-and-yellows. Joining them are eight blue-headed macaws, with their pearly eyes and two-toned tails, half a dozen chestnut-fronted macaws, two red-bellied macaws — it's quite a floor show.

The big red-and-green macaws are the heavies here and can be very assertive about claiming or sticking to a spot. They might stay on the clay as long as twenty minutes. Scarlets might stay half that long. The smaller blue-and-yellows are more likely to hit and run, staying about five minutes.

For those now on the face of the bluff, there is work to do. This is not wet clay. It must be chipped or chiseled. With much effort of their can-opener bills and thick tongues, the macaws gouge into the hardened slope, prying crumbles. These are their prizes.

The clay shows the efforts of years of such digging. Under the overhanging vegetation at the top of the wall, some birds disappear into hollows and little caves, the work of macaws past and present.

Some chip off a chunk and fly away with

what Don calls a "to-go box." They take their prizes to a nearby branch and, holding them like ice-cream cones, nibble leisurely. Later, in deep forest nests, sodium-rich clay will be in the food fed to their chicks.

As individuals, most of these birds come to this clay bluff about every other day. But even when they come, they don't all go on to the clay buffet on every visit. This is also the big social scene. And parrots are *very* social.

As people head to a bar "for a drink" when they are really going to socialize, there is more here than just clay and sodium. It's a place to see who's around, maybe who's up for pairing. Parrots *need* to socialize, and like us they spend time and energy to go places simply to be with others of their kind. It's worth it. In parrot cultures as in human cultures, socializing is important.

One of the most striking things about their lives is that they have a lot of free time on their wings. They never seem to be in a rush.

"It's not like they have an urgent need to find food," Don notes. Food is plentiful for them. Macaws can fly about thirty miles an hour, "but they don't move around much or go far." Don has studied their movements by affixing GPS trackers to several birds. He's found that they use the landscape at a pretty relaxed pace. Flowers and fruit are pretty easy to spot while they're flying, he says. "So the macaws can keep track of what's ripening, where, and when." They have ample time for socializing.

In places such as this, not only do they congregate in flocks; within those flocks they interact both as individuals and as pairs, like human couples in a group of friends. Relationships are maintained, developed, or begun.

Don points out two white-eyed parakeets sitting together, touching each other while facing outward. They're in a big flock. But as this one pair indicates, within the crowd relationships are forming like stars being born out of cosmic nebulae.

At first, each bird maintains a little sphere of personal space. Then one individual might begin to enter another's space. At first they might squabble a bit. But after a while, proximity inside the personal space is tolerated. Next they move from tolerating to interacting. And then they sit next to each other with feet touching, looking out at an angle like these white-eyed parrots we're watching. At this point, they're operating as a pair.

To human eyes, the process is "insanely subtle," Don says. It's another way that parrots differ from other birds. Many other birds have more ritualized courtship: specialized dances, synchronized flying, courtship feeding; bowerbirds even build elaborate courtship structures reminiscent of wedding bowers. By contrast, the parrots' free-form proximity really is more like a bar scene. Not even as ritualized as a high school dance.

Like young humans, the macaws indulge in certain fads. Sometimes for months they'll go almost exclusively to one section of the clay face. Suddenly they'll shift to a new spot for a few weeks. And then, maybe, they'll go back.

"It's like going to the hot club where everybody's going," Don says. "Then, suddenly, another club is hot. No real reason. No predicting it."

Oddly, macaws do not visit different clay licks. The famously photographed clay bank called Chuncho is only about six miles (ten kilometers) downriver from here. Any macaw here could fly there in fifteen minutes. But birds who go to one lick seldom — if ever — go to the other.

"There's complex behavior that I don't get," Don says. "A lot of times," he admits, "I'm just stumped. I think there's a lot going on at the cultural level."

Don tells me about one clay lick used by red-and-green macaws but never visited by the scarlet macaws who breed nearby. No red-bellied macaws or white-bellied parrots visit Chuncho. "I don't know why. It's very odd." He hears black-capped parakeets here, but they don't use this clay lick; they prefer a different one nearby. "Using this clay isn't in *their* culture," he says. "But others use it, so *why* don't they? There's behavior that doesn't make sense to me."

Actually, one aspect of culture is: it doesn't *have to* make sense. Colorado is the only clay lick that gets all these species, something Don says seems "arbitrary." But culture *is* often arbitrary. One human culture wears funny hats; another culture wears different funny hats. Our hats never look funny to us; they're just our hats.

Scientists like Don Brightsmith — and me, for that matter — make sense of the living world by first asking how and why something evolved; we trust logic, and we want to know the *reasons* animals do things. That line of questioning gets you far. But sometimes it doesn't get you all the way to understanding. Some animals seem to have evolved an ability to do things *for no particular reason.* Humans have this ability. Parrots seem also to have this ability. For culture to develop, someone must do something that has never been done. Then that thing must spread. Some traditions are just arbitrary habits that have spread.

"I get the feeling that parrots are creatures of habit," Don says. "For some reason, or maybe for no reason, or just out of their sociality, they'll start doing something." Birds watch one another, and they often copy. Sometimes, when parrots are in a flock and

306

one pair starts preening, many pairs in the flock will begin preening. "I think there's a lot of careful watching going on," he remarks.

The noise *feels* like communication. *Something* is being communicated. The parrots know what. Don says he's been in places where flocks of Amazon parrots have stayed quiet. The difference: there, people hunt parrots for the pot; they eat them. The parrots understand that they are saying something that can be heard, overheard, responded to, and even used against them. But when they make noise, it doesn't seem to be language like humans have, with our large vocabularies of specific words and rules of grammar and syntax.

Perhaps what they're doing with all their vocalizing is what humans do when we get together to play music; perhaps it simply builds cohesion, a sense of group identity. If sheer vocal expression is older than language, then perhaps art preceded words. If art is older than words, perhaps that is why the world is full of *beautiful* display. Maybe these birds are all artists, masters in the art of being parrots, playing their parrot jam sessions.

Growls, caws, gurgles, and screams. Some of the time, parrots remind me of ravens. One thing: they *are* all listening. Despite so much noise, an entire flock may suddenly go silent. Sometimes only the ones on the clay face will spook and the ones in the trees will stay put.

Or a particular scream will make everyone leave the theater for the rest of the day. There seems to be, within all the noise, a great deal of paying attention to each other, not just calling but *listening*.

Joanna Burger studies bird behavior and was my PhD adviser. I knew Joanna's adopted red-lored Amazon, Tiko, for many years. Tiko, who would often watch vigilantly through the windows, uttered different, recognizable alarms for "hawk" and "cat." Hatched into captivity, Tiko lived to be sixty-six years old. It's possible that his fears were innate and his words self-invented. Wild parrots, with long traditions, rich social interactions, continuous learning opportunities, and real imminent dangers, might well fine-tune such instinctive apprehensions and alarms into highly informative vocal and behavioral messages.

When Joanna said to Tiko, "I'm going to the garden" or "I'm going to the office," Tiko always flew correctly to either the office door or the garden windowsill. This suggests that parrots, with their complex vocabularies, can extract meaning. Or, at least, they can learn names of places.

For title of best vocal mimics in the other-than-human world, parrots have few rivals. But many captive parrots merely "parrot" human speech without seeming to understand that words can have meaning, can refer to something. Only humans are known to have language where grammar — parts of speech, the arrangement of words, inflection — allows for nearly infinite combinations. Language lets us refer to any conceivable event, thing, time, quality, idea, and so on. Non-human signals often seem limited to declarations — "I am here"; "This is mine"; "Give" — and to emotional states such as threat, nonthreatening intent, and danger alarms. Non-humans don't seem to talk about things outside of themselves, like the weather, sea conditions, or how today's food compares to yesterday's.

But now, here's an astonishing thing that's been discovered in our lifetime: other-than-humans across a range of beings from parrots to apes to dolphins have, in captive settings, learned to understand and even produce words and phrases that refer to objects and even some concepts. This opens two big doors: If the capacity for referential understanding is there, are we missing it in the wild? And if they aren't using this capacity in the wild, how and why do they have it? So far, we don't have the answers.

We do know that coaxing this ability into human view requires special training. Captive chimpanzees and bonobos can learn to use hundreds of symbols. At least one border collie, named Chaser, dubbed "the world's smartest dog," knew the specific names of more than one thousand toys and other objects and could fetch each when asked.

Some parrots have understood that nouns and numbers are abstract representations of real-world things. And they *have* used simple grammar rules to utter sentences and make requests. African grey parrots can learn to use dozens of human words and phrases to identify and ask for things. To break the species barrier with these parrots, a specific training technique is required. There must be two humans who demonstrate, who socially interact, and who take turns using the words to refer to and act on the things — items, shapes, colors — in question. As the parrot begins learning, the humans must adjust the pace of teaching. Without these techniques, the birds usually *don't* learn that certain words mean certain things.

During the learning period, the parrot will often practice in private. After a parrot thoroughly gets it, they will sometimes invent their own labels for things by joining words. Alex, a closely studied African grey parrot, combined words to refer to new things. For Alex, an apple was a "banberry," part banana, part cherry. That's as much as chimpanzees who've been taught to use sign language and symbols have done. Alex used the verb "go" to express requests ("Wanna go chair"), announce intent just before acting ("I'm gonna go away"), and demand that a trainer leave ("Go away").

What this shows is not that parrots can learn meanings of human words. It shows that parrot minds understand that some things *refer to* other things. We've recently discovered how they use their own vocalizations as names, and as group-identifying dialects. Their need to learn things socially — undergirded, of course, by their mental proclivity for social learning — hints at the profound importance for them of growing up in groups and absorbing (and perpetuating) the local culture.

Prairie dogs' alarms tell others what *color* the predator is. The remaining question about them, as with parrots, is: In their free-living cultures, what else might they be referring to, as they move through the vivid experience of their days and lives?

One pair of scarlets lands at the outer branches of a tall tree. These two start fooling around, hanging upside down, billing and gently biting, their wings flapping like flickering pilot lights in the leafy greenness. When socializing, macaws often hang upside down and goof off. It appears that parrots have a sense of humor, or fun; at the very least, they are quite capable of being in a good mood. We see two macaws actually copulate; and this is very strange. It's past mating season;

311

nesters already have chicks hatching. The flirty birds are probably young, bonded, sexually active non-breeders. They're like teenagers playing on a swing set — and one thing has led to another.

Just how well do these birds know one another? Don and his colleagues have seen groups of six or eight birds wearing leg bands applied by his team. That means they're likely young ones fledged from nests near each other, and that they know each other well enough to travel together as a group of young friends.

For macaws, particularly, their entire body is a flag of identity. Tracts of tiny facial feathers create unique patterns of stripes and dashes, like tattoos. Some scarlet macaws wear wide yellow shawls, while others sport narrow yellow shoulder scarves. In those fields of yellow, some bear dense blue dotting. Others, sparse.

Many territory-holding birds recognize their neighbors. Mockingbirds, for instance, don't fuss much with each other after their boundaries are settled but will aggressively drive off any new intruder. Hooded warblers recognize the song of their neighboring territory holder from year to year, even after migrating south, wintering, and migrating back north to their breeding territory. Seabirds such as terns, gulls, penguins, albatrosses, and others recognize their young among hundreds by voice and, in some species, scent. Various primates, crows, dogs, wolves, and horses recognize their group members by voice *and* on sight. (This is likely true of many social mammals.) Pigeons, crows, and jackdaws can recognize familiar individuals of *other* species. Our chickens, and a hand-reared orphan owl, are thoroughly at home with our dogs but flee in panic if a friend arrives with an unfamiliar dog. Birds are extremely sharp observers; they know acutely what's going on, who is in their world and in their presence.

The macaws play and spar, sometimes locking feet with one another and tumbling through the sky. On the clay, they sometimes show dominance by flashing open wings, showing off their bright underwing color. Bright color indicates health and vigor. It goes along with being in your prime of life. Being in the prime means you're a contender.

Most of what happens here at the clay is peaceable. But some birds parry with their beaks or give another a shove. There is no lack of clay or of space. "They're pretty peaceful. But they *do* fight occasionally," Don affirms.

About two hundred macaws use this particular clay lick. Social knowledge in the mind of each individual travels with them to wherever they might meet. To a nest site, perhaps — where much is at stake. Bonded

313

young-adult pairs who lack a nest site of their own will visit nests held by established pairs. They must decide, simply, whether to attack and carry out a prolonged fight for owner-ship of the nest. If they already know the nest-holding birds, part of the decision becomes easier. And safer. "From socializing at the lick," Don observes, "they probably know who is tough and who yields easily."

As they get acquainted with each other, sort out dominance, flirt, and bond, relationships made here can be carried with them as easily as a beakful of clay. Anyway, their dangling upside down, their flapping and fooling around, looks like fun. It looks like joy in liv-ing.

Some people would say I'm attributing hu-man emotions to non-humans, that I'm guilty of anthropomorphism. I'm not attributing. I'm observing. You can see things in parrots that you don't see in other birds. Parrots play, parrots mimic, parrots prance in rhythm with human music. Finches don't, and gulls don't. This is not anthropomorphism of parrots, nor attribution. It is understanding, based on observation.

The lower animals, like man, manifestly feel pleasure and pain, happiness and misery. Happiness is never better exhibited than by young animals, such as puppies, kittens, lambs, &c., when playing together, like our own children. . . . The fact that the lower animals are excited by the same emotions as ourselves is so well established, that it will not be necessary to weary the reader by many details.

— Charles Darwin, *The Descent of Man*

Only a few of the roughly ten thousand species of birds seem to play and fool around for the fun of it. The *most* playful of them are crows and parrots. Like playful mammals such as rodents, carnivores, and primates, playful birds chase, play-fight, and toss objects. Captive rooks will play tug-of-war with a strip of newspaper while they're standing in many strips of newspaper. Online videos show ravens using a piece of plastic for slidding down snowy roofs and car windshields, and cockatoos dancing to music having a heavy backbeat. In one online video, a group of swans surf the crest of a wave into the beach, then fly back out to catch another wave.

Play is practice for skills needed in life, but what *motivates* animals to play isn't their concern that someday they'll need the skills they're practicing. They play because the short-term reward is pleasure. Birds' and mammals' brains manufacture the same opioid and dopamine neurotransmitter chemicals that motivate pleasure seeking (more technically, "the search for reward-inducing stimuli") and create the feeling-good experience we call pleasure. Play is ultimately practical, but animals play because it feels good.

Play is also beautiful. Beauty is also pleasurable. Here's where the threads of the practical, the pleasurable, and the beautiful begin to braid. These feel like first drops of insight that I hope will gather into trickles of understanding and eventually flow into our comprehension of beauty as we seek to explain *why* macaws — and so much else — are so beautiful.

Another trickle toward understanding is that the (often beautiful) singing of birds isn't automatic. Birds interrupt their singing when they're startled or there's danger. In order to sing, birds have to feel motivated.

Of the varied parrots in the trees and on the ground, out of their polyglot noise, Don hears the juvenile call of a recently fledged blue-headed parrot. His listening abilities defy credulity. But there it is.

Singing — and whatever we can call all the noise the parrots are making — serves social functions. Among the most common functions are: staying in touch, attracting a mate, and proclaiming a territory. Many birds — and some mammals — sing. Song is mostly affirmation: "I, here, make this claim: Rivals, be forewarned that this place is taken. Neighbors, hear me stake these boundaries; know by my voice who I am. Ladies" — males do most of the singing — "come and feast your senses on my territory's splendor and my competent abilities."

If you could soar above and regard the landscape as birds sing it, you'd see polygons that exist on no map but stand emblazoned in birds' minds, asserted in calls and enforced at the boundary lines and hedgerows where the fighting is real enough.

And so a special footnote here for the humpback whale. Although birds sing to hold territory, humpback whales hold no territory. Birds sing to defend food supplies; humpback whales stop singing in the foraging season. Birds also sing to attract mates. Researchers have never seen a female humpback approach a singing male. Yet humpback whales sing the most complicated, unusual, and cultural non-human songs on Earth. For millions of years without pause, somewhere on this planet the ocean has hummed with their singing. What urgencies being conveyed are we so profoundly overlooking? Humpback whale song carries peculiar mysteries. The whales sing; that, we know. But the behavior during and following their singing is not consistent. When humans do complicated things, we proudly say it's because humans are complex. When other animals do complicated things, we feel frustrated that we haven't figured out "the" reason for the behavior. I think the evidence, conflicting and contradictory as it is (and precisely because it is conflicting and contradictory), suggests that humpback whales sing: to advertise their presence in the world, to socialize with other males, to compete for status with other males, and to impress females. Perhaps their changeable songs have various social functions for them, a bit like our own music.

Our own often-beautiful music incites and inspires different things. Sometimes, nothing at all. But no, that cannot be right, because there is more music in the world than we'd seem to need — we insist on making and hearing it. When I am writing, I often listen to music all day long. I enjoy how it makes me feel. It even seems to help me think. I don't know quite *why.* Roger Payne — whom we've met as co-discoverer of the song culture of humpback whales — tells me, "We feel it is so important to sing but have a hard time figuring out why we like to sing. I think that's because singing is so old. I think singing is older than humans, older than whales, older than the frontal cortex. Probably controlled by the lizard brain, where things unconscious take place." For that matter, singing is much older than lizards; various amphibians and fishes sing, and it evolved independently in insects. "But *why* is music so important?" I pressed, and Roger considered the question.

318

"We can't know," he said, "because it happens without going through words; it's direct emotional communication. If you're trying to make friends or mate or gain confidence, it would be wonderful to do it directly. Music does that."

Singing must feel rewarding, pleasurable; or else whales, humans, and birds would not keep singing. As we've mentioned, areas of birds' brains involved in singing include those dopamine and opioid receptors. But we didn't discuss how the two brain chemicals work in tandem. Researchers have found that dopamine — the motivating "feel good" hormone — provides the drive to sing and to keep singing when a female is interested. After mating, opiates reduce a male's motivation to sing. To simplify: dopamine produces pleasant, pleasure-seeking motivation; then opioids give a pleasure-attained sense of reward. It's a sort of "life is good" sensory perception system, one that humans basically share with all other animals.

So, birds sing to proclaim territory, to attract a mate, and because it feels good. Anything else? I asked David Rothenberg, the author of *Why Birds Sing* and *Nightingales in Berlin.* "Birds sing," he said, "because evolution is not just survival of the fittest, but also survival of the beautiful."

That seemingly offhand remark delivered a

very large conceptual can of worms.

What is beauty? Is it truth, as Keats wrote? Maybe it's just that simple. But —. Because we perceive as beautiful everything from truth to a whale's hummings to air forced through a bird's bill, deeper and harder questions step forward: Why do *we* perceive *their* songs as beautiful? And, indeed, *why* does the perception of beauty *exist*?

Clearly, I'm going to have to think a lot more about this.

Singing emerges fully after a process of social discovery. When we talked about vocal learning earlier, perhaps you wondered whether it's really as simple as "male chick learns song from Dad." You were right to wonder. Nothing's ever that simple. And there are always variations.

In cowbirds, young males learn to sing from females — who don't sing. Researchers based at Indiana University brought some cowbirds into a huge aviary. Every once in a while, a male would suddenly change the pace of his singing and abruptly move toward a female. What was up? The female was using a gestural signal — flicking her wings — to let the male know which of his songs she liked a lot. Males often repeated that favored song, practicing and memorizing what was working. Later, females assumed mating postures when they heard the songs that had prompted them to flick their wings. The researchers wrote that cowbird song learning wasn't a simple matter of a male copying an older male but "a process of social shaping." Males had to perform songs over and over in order for the females' reactions to become evident. Female reaction is what shaped male singing.

Moreover, females stayed to listen more often when they heard the singing of males who'd grown up in the same neighborhood they had. When a male from out of town sang to local females in the aviary, females often flew away even before the out-of-towner finished his song. It turns out, different populations have different song cultures. Different songs translated into differences in mating success. Wrote the researchers, "What impressed us most was how social behavior could serve as a variation generator." What that means is, individuals with similarities formed groups — and differences wedged the different social groups apart.

Cowbirds aren't unique in this regard. Competition and a preference by females for males who sing the local dialect has been documented in Galápagos finches, African indigobirds, and others; it's probably widespread.

So there's much more going on in their own vividly experienced social exchanges than meets the human eye and the ear. Songs, dialects, interactions, complexities — theirs are richly social lives. It's far from grim survival. The normal behavior of birds and so many other animals is often more diversified, and more reliant on socializing, than we assume. The cowbird researchers conclude, "Culture or traditions are learned as animals probe the properties of their social surroundings. Skills can come from imitation and teaching, but ultimately depend on social performance and feedback."

In many birds, females do the choosing. It appears that female choices can keep populations together or apart.

But wait. That line of logic leads to an interesting question: Could social groupings, driven by female preferences, cause enough divergence to eventually create new species?

Essentially all evolutionary biologists believe that the fork in the road to new species requires, first, that some group of individuals becomes geographically isolated by a physical barrier, such as a rising mountain range or a new river. Classic examples are the closely related different finches and tortoises on the different Galápagos islands. When isolated populations come under different pressures from changing environments over long enough stretches of time, the accumulating differences can make the two populations into separate species.

But I've long felt that geographic isolation can't be the *only* route, and that something else must also be at work. In some African lakes, cichlid fish have diverged into an astonishing array of species, seemingly without geographic isolation — because, after all, they're all stuck in the same lake. Is it possible that without geographic isolation, groups can form, and diverge, because of what amounts to purely cultural, even arbitrary, preferences that create both group identity and between-group avoidance? There's a lot of that in humans, obviously. And in at least some other species, such as killer whales. White snow geese tend to prefer white mates, and darker birds of the same species, called "blue" geese, tend to prefer blue mates. But could diverging cultural trajectories, given enough time, diverge *permanently*? Could species result without geographical isolation because of culture alone? Might cultural mating preferences based on something as insubstantial as song or dialect — or even beautiful colors — be a widespread driver in the origin of *thousands* of species? It seems a slightly crazy thought, and slightly heretical. As I've said, I'll have to think a lot more about this.

Suddenly all the birds on the clay flicker up into big, leafy trees. Someone shouted an alarm. The birds trust each other's judgment, no questions asked. Don says, "I don't know why they're so jumpy today."

Why: an orange-breasted falcon lands in

the top of a tall dead tree. The falcon's big feet indicate bird-killing capability. But there are just too many alert eyes here, ready to alarm at the first sign of danger. Despite the buffet selections spread out below, the falcon's next move is to vanish into the forest. She needs a better chance to surprise.

The macaws start draining away across the rainforest's emerald skyline, sparking the air with crimson embers.

In a few minutes it will be eight a.m. How long and rich a morning can be if you bring yourself fully to it. Come to a decent place. Bring nothing to tempt your attention away. Immerse in the timelessness of reality. Attention paid is repaid with interest.

Dominica rises from a deep blue sea that is home to sperm whales.

Jocasta makers her presence known.

Shane Gero listens for sperm whale clicks.

Electronic tag with suction cups

A neat bite scar

Shane wields the suction tag on a long pole.

Sperm whale tails are all different.

"… a sudden explosive leap and terrific crash"

The Tambopata River descends from the highlands of Peru's western Amazon region.

Macaws hatch blind and helpless.

"Like flying flames of red and yellow and blue"

Read-and-green, scarlet, and blue-and-yellow macaws along Peru's Tambopata River

Macaws of three species arriving at a clay "lick"

In the salt-starved western Amazon, macaws mine the sodium-rich clay of an ancient seafloor.

Gaby Vigo

Don Brightsmith

An ancient fig tree with root buttresses like ribbony garden walls dwarfs Carl Safina in the Amazon rainforest.

Ben

Alf

Cat Hobaiter observing free-living chimpanzees in Budongo Forest, Uganda

FROM LEFT: *Hawa, Musa, Simon*

Irene and her six-month-old infant, Ishe

Chimp showing peak estrus swelling

Lotty

Cat Hobaiter

Kizza Vincent

Masariki

Monika

Tense and silent, on patrol against their neighboring enemies

Nora and young Nalala

Talisker, the senior statesman

Pascal, drinking with a leaf-sponge

BEAUTY:
FIVE

About that conceptual can of worms: Rothenberg's remark, "Birds sing because evolution is not just survival of the fittest, but also survival of the beautiful" was a quiet little bomb tossed to me, and innocently I caught it. And now the quietly explosive implications are beginning to worm their way in.

Macaws are unusual. In most birds, one sex is more colorful and vocal, usually the males. But with macaws, both sexes look and sound similar. Compared to the dozens of leaf-green, well-camouflaged parrot species, the three main macaw species here, flaunting and flickering in the treetops, are entirely obvious. These macaws are so unabashedly spectacular that they are, for all practical purposes, nakedly beautiful.

Why?

The spectral bat, *Vampyrum spectrum,* kills and eats birds the size of Amazon parrots. But red-and-green macaws are three feet long, followed closely by scarlet macaws and the only slightly smaller blue-and-yellow macaws. No predatory bat or raptorial bird is big or stealthy enough to eat these macaws regularly. Having escaped most predation by sheer scaling of size and keen-eyed crowd-sourced vigilance, macaws have broken out of the basic green camo fatigues common to being a parrot in the New World. It's as if they have freed themselves to let beauty evolve according to the whims of aesthetics.

But is that possible? Could "survival of the beautiful" *explain* macaws? Could beauty itself *evolve* through socially leaned preferences, through *culture*? The question itself seems almost nonsensical. But look at these birds: the impression is inescapable.

Let's first consider a species in which the male alone looks amazing. A peacock will do nicely. The female, the pea*hen,* is of muted hues, camouflaged. Charles Darwin's theory of natural selection could easily account for this camouflage: obvious birds would get eaten while well-hidden ones survived and proliferated.

Darwin's great principle, so sufficient to explain the hen, could not possibly account for the peacock's hazardous extravagance. In fact, natural selection failed to explain a lot of the lavishness of the natural world. Darwin himself knew it. At a loss for what else was needed, Darwin confided his perplexed frustration in a letter to botanist Asa Gray, writing, "The sight of a feather in a peacock's tail, whenever I gaze at it, makes me sick!" The Duke of Argyll demanded, in exasperation, "What explanation does the law of natural selection give" for decorations such as tail spots on a hummingbird?, then concluded, indignantly, "None whatever." Darwin commented, "I quite agree with him."

Something was producing breathtakingly emblazoned, highly impractical body decorations across the living world. Look online for photos of racket-tailed hummingbirds, and you'll get one idea — among innumerable examples — of what was making Darwin lose sleep. He'd launched his great and logical argument for natural selection. That ship had sailed. But even in the pages of his towering *On the Origin of Species,* he'd acknowledged a serious leak. Natural "selection" is in fact a misnomer. The environment does not seek, choose, or actively select for anything; it filters things out. If your color pattern makes you obvious on a nest, you get filtered out. Natural selection's harsh live-or-die struggle for existence just did not seem capable of creating the world's abundance of wildly extravagant plumes, bright colors, ornamenting horns, etched markings, and intricately diverse songs. What could be having so major a role in driving such proliferated gorgeousness?

Much of the wanton decoration was wielded by males. From that, Darwin took his cue. "With the great majority of animals," Darwin could not help noticing, "the taste for the beautiful is confined to the attractions of the opposite sex."

Perhaps you have heard of bowerbirds. About twenty species, dove- to crow-sized, live in New Guinea and Australia. The males' "bowers" are the most elaborate things built by non-human beings. Some species build roofed, hutlike structures; others create walled lanes. Thousands of twigs and stems go into the construction. In and around these, the male decorates, using hundreds of collected objects.

The bower is not a nest. It is only this: a seduction theater. Its sole function: to convince skeptical females that they should want to mate with the builder. Bowerbirds appear to be the only animals other than humans who use found objects as aesthetic lures of seducement.

As the bowers differ from species to species, so do their decorations — flower petals, shells, feathers —. Certain species prefer blue objects, or red or green or whitish ones. In some species, he uses a tool and some berry juice to paint color onto his walls. Nowadays they incorporate colored glass, nails, and plastic items. The male spends hours carefully arranging these. Some create optical illusions of structural size by ordering the items from largest to smallest to produce a forced perspective. Move something or deposit an item of the wrong color, and the male will likely fix the display, returning it to his preferred arrangement. The sole consideration appears to be visual effect, how things look, a conscious aesthetic regard.

Now a female comes, inspecting. The male manipulates some of his decorative items, showing them off like a salesman. She evaluates. Each female circulates among different males' bowers, which vary in impressiveness. Of course they do; that's the point.

Young females focus their evaluative attentions on the physical structure the male has built. More experienced females scrutinize his dancing and listen carefully to his calling. But also, the most mature females visit fewer males. They have apparently already met the local guys, and know whom they consider attractive. (After several especially attractive males in one population died, older females visited a larger number of males, checking out those who were newer, younger, or previously passed up.)

If a male who is dancing and calling for a female notices that she seems nervous, he will tone down his song and dance. If she seems increasingly interested, he will pump up the intensity of his efforts. She, based on his bower, his appearance, and his performance, will choose a mate.

Their relationship is only sexual. The only thing she gets from a male is the stimulation he and his bower spark in her — and if it pleases, she will obtain for herself his semen and its DNA-containing sperm. The male plays no other function, neither nest building nor parenting. He is a pretty boy, a busker, a hardworking installation artist, trying to impress for sex.

To *learn how* to impress, young males spend years perfecting their craft. Satin bowerbird males require four to seven years to learn how to make "competition-quality" bowers. They visit bowers of adult males. In some species, adults even let young "apprentice" males help them build their bowers. The youngsters are learning what a good bower should look like, not only for their species, but also in the specific cultural traditions of their particular population.

The competition is intense; most males strike out, while a few males get a lot of comers. After mating, the female vanishes to build an inconspicuous nest and raises the young by herself.

The strange beauty of all this confounds some scientists; they tie themselves in knots trying to explain why such extravagance exists. In science, the rule of thumb is that the simplest explanation is preferred. Two rules apply: Occam's razor and Morgan's canon.

Occam says that we should choose the simplest explanation for any observation. Morgan directs us to select the explanation for an animal's behavior that attributes the least and lowest mental functioning. (Try this for people; it works well.)

Where does this leave us with regard to male bowerbirds? Proffered explanations are, at best, a reach. Some scientists have suggested that the bowers' showiness allows males to be less colorful, and so less conspicuous to predators. That's wrong; some bowerbird males are stunningly showy, such as the almost Day-Glo flame bowerbird. Others hypothesize that a good bower indicates a male with a low load of external parasites. (Wait — what?) Another hypothesis is that the bower helps protect females from forced copulations (somehow). If any of that were realistic, most birds in many families would be building bowers. None of these mental contortions seem to explain why there are bowers, why styles vary, why some birds strongly prefer decorations of red, others blue. Occam and Morgan lead us astray. Bowerbirds take years to learn their craft (simple instinct doesn't explain their exertions), adjust their dancing to a female's responses (low mental functioning doesn't explain their acumen), and are aesthetically fussy (because bowerbird females are exceptionally judgmental).

Try this for the simplest explanation: Females simply find males and their bowers *beautiful.* Is that scientific? Charles Darwin thought so:

> Males display their gorgeous plumage and perform strange antics before the females, which standing by as spectators, at last choose the most attractive partner. . . . I can see no good reason to doubt that female birds, by selecting, during thousands of generations, the most melodious or beautiful males, according to their standard of beauty, might produce a marked effect.

If only beautiful males get to breed "according to [females'] standard of beauty," we'd see many species in which males are very beautiful.

We do.

So Darwin proposed another mechanism of evolution to complement natural selection: "sexual selection." Boldly, Darwin wrote of the world-altering power of sheer aesthetics, of beauty "for beauty's sake":

> A great number of male animals, as all our most gorgeous birds, some fishes, reptiles, and mammals, and a host of magnificently coloured butterflies . . . have been rendered beautiful for beauty's sake; but this has been effected . . . through sexual selection,

that is, by the more beautiful males having been continually preferred by the females, and not for the delight of man.

Darwin could find no alternative: "If female birds had been incapable of appreciating the beautiful colours, the ornaments, and voices of their male partners, all the labour and anxiety by the latter in displaying their charms before the females would have been thrown away; and this is impossible."

It seems impossible that humans alone appreciate beauties that are tuned to bring birds together, and that birds themselves do not. Logic precludes it. Moreover, their passions, their ardors, likewise preclude it.

Did Darwin correctly perceive male "labour and anxiety"? Singing zebra finch males who have no audience show brain activity in the regions involved in vocal control, song learning, and self-monitoring. When a female is listening, though, the male's learning and monitoring activity stops. It appears that the bird's concept of what he is doing changes when he has an audience. Now the male is *performing* for a female who is *evaluating*. It's like the differences between practicing an instrument and playing music. The song isn't just a program that the brain executes by pressing a Play button.

Bottom line: Darwin thought females simply prefer what they prefer. Darwin was convinced that sexual beauty is its own reward. Riffing on a pheasant's tail, he wrote: "The most refined beauty may serve as a sexual charm, and for no other purpose." And on this basis of pure and arbitrary aesthetics, females indulge their fancy for sex.

This was too much for natural selection's co-discoverer, Alfred Russel Wallace. He could not imagine that such beauty could be so *arbitrary,* that it had no *basis.* Wallace energetically asserted that female choice *had* to be utilitarian. "The only way in which we can account for the observed facts," he contended, "is by supposing that colour and ornament are strictly correlated with health, vigor, and general fitness to survive." For Wallace, if a bird preferred shiny feathers in a mate, it was because shininess revealed health and fitness.

And that's true, too. After all, dull feathers would betray poor health and nutrition. A beautiful prospective mate is one who should be a good breeding partner. Choosing mates having worn-looking feathers, dull hair, aged skin, low energy, and so on, isn't the best ticket to plentiful offspring. Attraction to those attributes would dead-end.

Further, much courtship is highly practical. Male terns often strut with a fish in their bill; a female accepts the offering before she will mate. ("First, take me to dinner.") Male ospreys do sky dances while dangling *their* fresh-caught fish from their feet. ("Look — I'm a good provider!")

But there are many animals in which courtship does not involve feeding and the males are very obviously extravagant. Why does a male red-winged blackbird have those red patches or a male yellow-headed blackbird its golden hood? That's pretty arbitrary decoration. At face value, the specific pattern and particular colors — they're meaningless. "If eyes," wrote Ralph Waldo Emerson, "were made for seeing, / Then beauty is its own excuse for being."

Splitting the difference between Darwin's beauty "for beauty's sake" and Wallace's idea that ornament must be "strictly correlated with health," the twentieth-century theoretician R. A. Fisher offered a middle path. He believed that female selection of males could start with, say, a simple preference for indicators of health — brighter feathers, perhaps, or a dash more color, or a vigorous song. Let's stick with feathers. In a competitive world, sons inherit the attributes of their successful fathers. Over successive generations, what counts as bright enough to be chosen becomes brighter, shinier, showier. At this point, bright feathers no longer merely reflect health. And mere health is no longer good enough. Now you need extremely bright and showy plumes to get chosen. When a certain look starts to be *the* look that females prefer, either you have it or you get left out.

It's as if a style or a fad begins to biologically evolve. Simple choice can push the preferred criteria to extremes. Fisher called this "runaway selection." The trait has lost its meaning and become an arbitrary requirement.

Because Life offers more examples of everything than anything we can imagine, there are plenty of situations where Darwin was right about the arbitrariness of aesthetics, Wallace was right about their utility, and Fisher was right about how they relate and diverge. Fisher's middle path may well be the route to Darwin's arbitrary extremes.

As the look of choice (or the song, or the dance) becomes more extreme, it becomes its own criteria — whatever "it" is for a bowerbird or a macaw or a human. What is perceived as beautiful becomes "it." And "it" becomes what is perceived as beautiful.

What does *not* seem possible is that a bird could exert such singing, dancing, fighting, or choosing and be wholly devoid of felt experience or sensual pleasure. The macaws aren't hanging upside down while flapping and squawking with their friends for nothing.

For Richard Prum at Yale University, the physical feather is a "signal." The perception of it as beautiful is an "evaluation." Long plumes are simply long. "Beautiful" is not a property of the thing itself. Beautiful is a sense created in a mind. A *sense* of "beautiful" is an evolved capacity of minds. Beauty is a kind of translation, the way our perception of electromagnetic radiation entering the eye at a wavelength of 680 nanometers appears to us as red. Our mind alone makes it red. Some things strike us as red; some things strike us as beautiful. How we form an impression of beauty is a mystery brewed deep into our nerves and glands.

Our brains give us an evaluation. And they give us an impulse: "Choose beautiful." We prefer smooth skin and lustrous hair not because they are inherently more beautiful but because they reflect health and youth, promising safe contact and perhaps a successful outcome to pairing.

But in the moment, we don't care about any of the rationale; we only feel the impulses. It's sheer aesthetics — the quality and brightness of the feathers, the alertness and ease of the dancer, the purity and complexity of the singing — that the potential mate (whether bowerbird or undergrad) is evaluating. They are not analyzing blood samples or taking biopsies. They are using an aesthetic evaluation that either attracts them or prompts them to move on. We don't have a multibillion-dollar hair-product industry to advertise that we are evolutionarily fit for

337

breeding; we have it because people like nice hair. Lotions and moisturizers fly off shelves because having smooth skin looks and feels good, that's all. Whatever the deep evolutionary reason, superficial impulses drive our behaviors. Looks are only skin-deep — but often that's all we care about.

And so it is for males and females across the living spectrum.

BEAUTY:
SIX

At this point in our explorations, two clarifying points need our extra attention. One is a question: Why do males so often compete and *females* do the choosing? It's largely because many kinds of males are mainly selling sperm, and sperm is cheap. Sex is a buyer's market. Females, who hold more precious goods, get to browse and choose. This general truism holds even for many birds in which males provide parental care, a role of greatly added value compared to mere hit-and-run fertilization. Females benefit from pairing with high-status males in either case, because the female investment in eggs and young is much greater and, therefore, riskier than male investment in sperm. So females require that males bring quality to the negotiating arena.

The second point is the real big one. Animals are often attracted to what they see other animals attracted to. This means that what an animal sees as attractive is *also* subject to cultural influence and social learning. If that seems subtle, don't be fooled. It's the big show-stopping dance number, with implications that reverberate across Life, through time, and to the far horizons.

It's obviously true for us humans that we're attracted to what we see others attracted to. Turns out, the social power of mere preference is astonishingly widespread. Female guppies like brightly colored males, but they can learn to like drab males if they observe many females mating with them. Zebra finches, small birds that are often closely studied because they adjust well to generations of captivity, provide further examples. If a virgin female zebra finch sees a female mating with a male wearing a white leg band, she will then prefer to mate with, yes, a male wearing a white leg band. (Or a red one, or an artificially applied crest feather with one particular stripe or another.) This tendency to "mate with someone who looks like one you've seen others mate with" has been documented in fish called mollies; it's even been documented in fruit flies. In carefully controlled experiments, female flies preferred either green or pink males, depending on which sort they'd seen another fly mate with. I'll quote the researchers, because you can't make this up: "Fruit flies have five cognitive capacities that enable them to transmit mating preferences culturally across generations, potentially fostering persistent traditions (the main marker of culture)." The process was one of "copying and conformity." The researchers conclude, "Culture and copying may be much more widespread across the animal kingdom than previously believed."

Runaway intensification of sexual advertising may hit a limit only when the advertisements — tail feathers, say, or antlers — become too great a liability. Presumably the peacock cannot afford a heavier tail. If you search out pictures of the now-extinct "Irish" elk (they lived all across Eurasia to what's now China), you can get a glimpse of what looks like fatally extreme antlers.

Among the dozens of species of North American warblers, females look inconspicuous year-round, while males become gorgeously colorized during the breeding season. And that means, once again, that females are in charge of choosing and they choose the beautiful. There are the black-throated blue warbler, the black-throated green, the black-burnian; the yellow, the yellow-rumped, and the yellow-throated; the redstarts, the red-faced, and dozens of others. One gets the impression that they are simply color-coded so that females and bird-watchers can (with practice) tell them apart. Why are the hummingbirds and the sunbirds and many other so flagrantly bejeweled?

The females don't need the males to be so colorful to make sure they're going to pair with the correct species. The males' differing songs provide sufficient, reliable species identification, even to human ears. If females *merely* needed reliable ways to identify species so that they don't waste time approaching the wrong potential mates, males could just sing their ID into the breeze, while looking as protectively plain as females. Or male markings could be as austere as barcodes.

Far from it! Their markings and hues go much further than mere identification would require. They are loudly lit, proudly worn, and shamelessly advertised. Lavish. Dazzling! Because the underlying driver is the females' choice, it is hard to imagine that they don't look fabulous to one another. Why else would they have such bright, strikingly detailed, wild-looking designs? Why else would their decorations and distinctions work *so* hard?

In large measure, the animal world is a female's world. Male beauty is the result of millions of generations of female selectivity. Males go courting because they have to. A hen prefers the showiest rooster. She chooses. Darwin was the first to see that. But Darwin then jotted a deeper question: How? "How does Hen determine which most beautiful cock, which best singer?" He filed this jotting under "Old & useless notes about the moral sense & some metaphysical points." Darwin seemed to have little patience for metaphysics. He suspected that all the answers to Life were in life itself. The evidence might be hidden, but clues lie all around us. "He who understands [a] baboon would do more towards metaphysics than Locke," he wrote almost defiantly. Get out of your head, he seemed to exhort; come have a good look around!

"How does Hen determine?" How indeed. I think we've identified pieces of the puzzle, forming part of the picture. Darwin's question about "which best singer?" now has some answers. For one thing, slowed recordings reveal a lot more intricacy in birdsong than what humans hear.

Making faster and more complicated sounds and dance moves is a sign of male vigor, quality, dominance, and desirability. For instance, a shiny blackish New Zealand bird sporting white puffs of feathers at the throat, called the tui, sings songs ranging from simple to complex. When researchers played recordings of a tui singing a complex song, resident males approached the speaker faster and more closely than when they heard a simpler recorded song, then immediately sang songs of higher complexity. They were more aggressive — perhaps angrier — and they used their own songs competitively. This sort of interaction reminds me of jazz jam sessions where the playing is so fierce and competitive that they're also called "cutting sessions."

As always with males, the females in the audience drive the whole system. Nearly 60 percent of tui chicks are fathered by males *outside* the pair's territory. In other words, more than half the time, the mother's mate — who is feeding her chicks and defending their territory — is not the father of those chicks in his nest. In other songbirds, too, males who sing more complex songs lure other males' mates into their own territory for trysts. With females shopping around and exerting so much discretion over their own indiscretions, males have good reason to feel threatened by males who sing better.

As Jennifer Ackerman, the author of *The Genius of Birds*, puts it, "Extravagance in nature is so often found in proximity to sex." There is no ducking the fact that female appreciation of the beautiful, even if it costs males dearly in energy and time, drives much of the world's courtship, and sex itself.

Michael J. Ryan at the University of Texas and Richard Prum at Yale have each thought and written a lot about why a sense of the beautiful exists. When I got a chance to speak to Prum about it, he told me, with resonant simplicity, "When females choose, they choose beauty. Beauty is the *result* of choice."

Is it always the result of *female* choice? Not always. In many species, yes, it's the males

344

who've got the showy outfit, the song and dance, and the females judge and select. But exceptions exist, many, even in birds. Macaw males and females, for instance, are equally flamboyant. Mockingbirds are equally muted, and both sexes sing. In many fish species, the males are more lavish; in many others, the sexes look similarly striking or the difference are subtle. In many mammals, the males are larger and have features such as antlers or larger horns, beards and mustaches, ruffs and manes. In many others, the sexes are equally marked (giraffes, cheetahs) or drab (rodents and bats). Male and female cranes look similarly graceful and albatrosses similarly face-painted, and both sexes dance equally. In humans, both sexes' bodies advertise maturity; both sexes are vain and careful about their looks and are usually choosy about long-term mates. Male humans are selling not just fertilization but the potential for reliable decades-long childcare and long-term bonding, making the best of them more valuable, putting them in a bargaining position to do some of the choosing. The result is that men are in competition with men and women are competitive with women. Of course, in humans looks alone still count for too much.

In species with intense male competition, males recognize and evaluate who they're up against. A male bird steps it up if his rival is singing unusually well. All of this implies that while female choice drives much in nature, the ability to perceive beauty is probably more widely and evenly distributed in both sexes.

The *capacity* to perceive something *as* beautiful, so as to make a choice, is ancient. The necessary neuronal connections and hormones were fully operational long before anyone who might be called human started glancing around to see who looked good.

Natural selection has been called Darwin's dangerous idea. Prum calls sexual selection "Darwin's *really* dangerous idea." Sexual selection leads to the conclusion that females, by the power of their mate choosing, have created essentially all of the beauty in the animal world. Living beauty is largely the manifestation of millions of years of female imaginations exerting arbitrary preferences.

That's enough to rock the planet. But let's venture much further.

Recall the cowbirds who will mate only with males singing the local dialect. Female cow-birds' choices reinforce cultural dialect differences, and that helps to both define *and separate* whole groups. We've seen how orca whales of the Pacific Northwest maintain groups that segregate themselves from others based on vocal dialects, how they never mingle, and how the mammal-hunting type has already evolved more robust jaws than the fish-hunting orcas. We've seen that young zebra finches and even fruit flies copy the arbitrary sexual tastes of other individuals, choosing for themselves the kind of individual they've seen others choose. Those are just a few examples of socially learned preferences causing the formation of groups of individuals that stay and operate among their own sort. This learned stickiness within the group produces the side effect of groups avoiding other groups.

I've called all this a can of worms. We've barely popped the lid. Taken to its logical end, these arbitrary preferences, if they spread culturally, generating groups that avoid each other, could drive a very major consequence: creating new species. Earlier when I was at the clay lick, and I first discussed female cowbirds' discriminating responses to dialects, I called this a fairly crazy thought. It's rather heretical. That doesn't mean it's wrong.

Darwin identified sexual selection as the driving force behind extravagant markings and plumes. But neither he nor the many others who've looked into it, from Wallace to Fisher to, more recently, the academicians Ryan and Prum, get as far as speculating that female preference alone could create new species. As I mentioned earlier, virtually all evolutionary biologists believe that essentially the only way a species can split into new species is, first, that populations become geographically isolated from each other, like the famous Galápagos finch populations cut off from each other on their different islands.

The twofold problem is: some members of a species must stop interbreeding with other members. Then, once these groups are isolated, differing pressures must shape each isolated population so that over time and generations, the populations accumulate enough differences to make them unable to breed freely if they meet again (that's the common definition of separate species).

But how do they get isolated? As mentioned, the usually invoked solution to the twofold problem is: geography. Either new mountains rise, new rivers form, continents drift apart, or some individuals reach an island of no return. This has been the answer since Charles Darwin pondered how and why various islands in the Galápagos archipelago each had different, though closely related, types of finches and tortoises. Geographic isolation works in theory, and it's been observed in reality. It is certainly a major stage set on which natural selection plays out. Because geographic isolation solves the twofold problem, few biologists have believed that one population can *ever* split into two species while occupying the same geographic region.

Known exceptions to geographic isolation are so few they have largely been regarded almost as oddities. Under certain circumstances, hybrids — offspring whose parents are of two different species — can facilitate the formation of a new species. Another mechanism for reproductive isolation without geographic isolation is called "learned imprinting." For instance, in the fly *Rhagoletis pomonella,* the adult lays eggs on a tree of the same species as the one it itself had hatched on. The introduction of apple trees to North America caused this hawthorn-tree specialist to split into two populations — one that lays on hawthorns and one that lays on apples. These populations then began diverging toward two forms. (Returning to lays eggs on the kind of tree they hatched on is learned individually, not socially, so it's not cultural.)

So: solving the twofold problem fundamentally requires some way to prevent two groups of the same species from interbreeding. But it doesn't fundamentally require geography to be the separating factor.

When I was a student, I harbored the heresy that populations remaining in the same region must have fractured into different species many, many times. But — how? A good place to look for evidence is a pond or lake, because inhabitants such as fishes have nowhere else to go. The existence of hundreds of species of cichlid fishes in the same African lake has always seemed to me proof that some other process is functioning in the world. Proof, yet that mystery remained: What process? Some experts believe that, even there, the fish must somehow *first* separate by occupying different areas — deep, shallow, shoreline, sandy bottoms, and so on — *before* they can evolve toward separate species. And this is plausible. A single large lake can have varied places, such as rocky shores or sandy or muddy bottoms or open water. Such varied habitats are called "microhabitats."

But, again, there are hundreds of species of cichlids in some lakes; it's unlikely that there are hundreds of kinds of places to get away from everyone. Something else must be going on. Les Kaufman of Boston University has studied cichlid fishes for decades. He told me, "Environments with many microhabitats give major advantages to individuals specializing in making a living in a particular microhabitat." The key question becomes: Do specialists actively *avoid* mating with fish that are still generalists and with others specializing in using the lake differently?

The same weekend I wrote to Kaufman on this subject, I phoned my friend Melanie Stiassny. Melanie happens to be the American Museum of Natural History's curator of fishes *and* a world-class cichlid expert. I told her I wanted to run an idea by her: Can behavioral specialties that are learned socially and passed along culturally create specialist groups that avoid other groups, causing them to separately evolve deeper into their specializations over time, eventually becoming separate species without ever having been physically isolated from one another?

Melanie wasn't too enthused. She said she was not a big fan of the idea that a species can evolve new species in the same place. She believes that cichlids in a place as vast as Lake Victoria, for instance, first became segregated by the lake's different habitats and *then* started evolving into different species.

But, I pressed, how would they first segregate? Couldn't culture alone do it?

Stiassny was still skeptical about the culture part. She promised to think about it.

So far, scientists have documented about a hundred species, from mammals to fishes to butterflies, in which individuals or groups use different specialized skills. But my question remained. So I dug deeper into the scientific literature, asking whether specialization spread by culture could have anything to do with the evolution of new species.

Turned out: it could. In some lakes, sunfish have developed two types of specialists within the same species. Moreover, their behavioral specializations have led to physical differences between the specialists. Open-water specialists have developed slightly longer bodies

351

more suited to travel, while along-shore bottom foragers have larger fins that are better for hovering in one spot.

Think again, for instance, of the killer whale types who inhabit the same region but specialize in hunting different prey in different ways — catching fish in one case, mammals in another — and have consequently developed social and physical differences. Regardless of the fact that scientists haven't named these whale groups differently (yet), they avoid each other and really have become separate species.

How might *cultural* specialization carry into *genetically* evolved differences? Imagine a fish species in which two kinds of specialists develop. Let's say that one group specializes in bottom-feeding, another specializes in open-water feeding. Let's say that young fish learn how to feed by watching older fish. And let's say there's an advantage to specializing such that specialists forage more efficiently than generalists, so they survive better. Relatively inefficient generalists, at a disadvantage, would begin to dwindle. Young fish that fail to specialize would not survive as well. Higher survival of specialists, and of young fish who tend to adopt the specialties of nearby adults, would push the two specialist groups apart. Over many hundreds of generations, specialists would start to diverge from one another. Specialists would begin evolving different shapes and behaviors that further improve the efficiencies of their specialty. Eventually they'd be different enough — maybe — to be different species.

This is no longer hypothetical. In ponds in the United States, a study found, 70 percent of bluegill sunfish sorted themselves into bottom or open-water specialists. In a different lake, pumpkinseed sunfish had also differentiated into bottom and open-water forms. Specialists got fatter and grew faster; generalists captured less food. That means there is a survival advantage to specializing. And this can drive genetic evolution.

Along Canada's west coast, fish called sticklebacks moved into various coastal lakes when the glaciers retreated at the end of the Pleistocene. In each of those lakes, independently, the original stickleback type evolved into two: a slender, open-water plankton-eating form and a shoreline-lurking stalker of crunchy invertebrates. The two forms in each lake avoid mating with each other and have become genetically differentiated, separate species.

In a lake in Nicaragua, scientists documented how fishes called Midas cichlids split into two species. First the fish began using different resources in the lake (some feeding along the bottom and some in open water, eating different things). Then the bottom and the open-water feeders began avoiding each other when it came to mating. The same thing has happened with cichlids in Cameroon and, indeed, with other freshwater fishes.

So far we have seen that over thousands of years, some diverging specialists turned into different species in the same lakes. Researchers have begun to realize that "specialization is widespread but underappreciated."

But can social learning alone cause specialists to be attracted only to similar specialists? Could a species first begin to break into diverging groups while occupying the same place simply because individuals learn and do what they see others doing, and mate with the types they grow up around and have seen others mating with? Culture could not create species unless the answer to those questions was: yes.

A few days after our phone conversation, Melanie Stiassny, having thought about the question of culture, as promised, e-mailed me. She wrote, "If any fish can be thought of as having a 'culture,' it would be cichlids. Very unusually for fishes, they have protracted parental care (both parents are involved with 'child care,' usually for a long time), so the opportunities for behavioral imprinting are particularly high. It makes sense that small preferences for habitat or parental coloration could be imprinted on offspring, leading to differential mating, ergo — speciation."

And, in fact, experiments with Lake Victoria cichlid fishes show that young females develop sexual preferences for males who look like their fathers — even when researchers rigged it so that the fish the females saw as their fathers were actually males of a different species. In a similar experiment with birds, for their entire lives the duped birds acted and reacted as though they were members of the different species who raised them. Other studies also show that behavioral specializations can be learned *socially.*

Meanwhile, I'd written back to Kaufman asking, "Do you see a role for groups diverging because individuals *learn* a specialization *from others* who've already learned it?" He answered hurriedly, "Waiting to fly from Nairobi to Kisumu for three weeks of cichlid study, seeing old friends, helping a Kenyan grad student. Answer to your Q: Yes. Of course."

We have seen from just a superficial survey of examples that in species ranging from sperm whales to birds to fishes to fruit flies, groups form and stick together and isolate from other groups of their kind based on socially learned cultural habits — habits as practical as foraging specialties and as seemingly arbitrary as differing preferences for certain dialects, colors, and markings.

Darwin, Fisher, and others have pointed out how arbitrary mating preferences result in peacock tails and elk antlers. I am now convinced that breeding preferences don't just create mating winners and losers who score high or low while playing for the same team — but that breeding preferences can and do create different leagues that take their own games into differing arenas. We've seen that cultural innovation and social learning create specialists. Once specialists occur, the stage is set for each to avoid others, take their specialization into a new niche, and permanently cut ties.

I strongly suspect that the mechanisms driving the origin of new species are mainly three: Charles Darwin's "natural selection" and his "sexual selection," and the one our present exploration has brought us to here, which I'll call *cultural selection.* By cultural selection I mean the power of socially learned preferences to create group cohesion *and* cause avoidance between groups. The avoidance means reproductive isolation. The reproductive isolation sets groups up for different journeys. Cultural learning can cause groups to mate like with like, thus deepening specializations, amplifying differences, and, I believe, diverging until they are sufficiently distinct to be different species.

The hundred-plus species of New World wood warblers evolved mostly in Central America. It strains credulity to think that nature would have commissioned Jackson

Pollock and Pablo Picasso, Philip Glass and Steve Reich to design so many closely related small birds in all their wild profligate proliferation of colorization and polyglot songs if the origin of each species had to literally wait for the moving of mountains and shifting of climates to isolate two populations. Nor does it seem that such physical features could actually segregate birds who annually migrate across continents and over wide seas and mingle in the treetops. It strains credulity to imagine that of the hundred-plus species in the kaleidoscope of fluttering butterflyfishes on coral reefs, each one had for thousands of years found itself utterly alone in an undisclosed location, then returned to the crowd fully transformed. That's not to mention the impossible-to-envision possibility that one clique of triggerfish could have found itself physically isolated from all other triggerfishes for the hundreds of millennia necessary to develop the bizarre and arbitrary patterning *ad absurdum* that earned it its name: Picassofish, *Rhinecanthus aculeatus*.

No. I don't believe so. I believe that without the necessity of waiting in each and every case for ponderous epochs to create geographic isolation, something pushed these creatures apart and sharpened their differences while they were poking around the same places. And that this something was: mating preferences that were *observed,*

learned, and copied, creating cultural specializations that bestowed advantages, driving
separation and ultimately sparking new species.

  Even the Galápagos Islands, the classic
discovery theater for geographic isolation in
the origin of species, is becoming the proving
ground for reproductive isolation by *socially
learned preference.* It is turning out that
Galápagos finches prefer mating with birds
that resemble their parents in looks and song.
Strictly as a matter of learning and preference, this means that for them "like mate
with like," which helps isolate them from
other closely related finches in the vicinity —
even of the same species. In the recent words
of Peter and Rosemary Grant, who've worked
painstakingly over four decades to understand
Darwin's finches, "Mate preferences that
develop from sexual imprinting on parental
body size and beak traits and from learning
paternal song . . . tend to maintain the
reproductive isolation." As evidence of cultural selection, the Grants' work, showing
that young finches *socially* learn sexual
preferences that determine who they'll mate
with, is a smoking gun.

  Sexual tastes and preferences — many of
them cultural, many of them female — have
helped drive Life's diversification. Likely they
drive it far enough to repeatedly cause the
origin of *beautiful* new species. Beauty — for

358

the sake of beauty *alone* — is a powerful, fundamental, evolutionary force. Beauty coupled with behavioral specializations, all reinforced by cultural learning that makes the young prefer the preferences of their elders, drives much of what we see in the wondrous living world.

BEAUTY:
SEVEN

Macaws are not only alike, male and female, but brilliant. Plumes, colors, patterns, personality — macaws have it all. And what we *see* — is beautiful.

Being big enough and smart enough to essentially avoid predation, macaws can afford indulging their own tastes for splashy tropical-fruit-punch colors and long-tailed elegance. Their beauty likely results from millennia of mates choosing gorgeous mates. But whatever the origins, it's stunning to realize that what *they* deem beautiful — *we* deem beautiful. Mysteriously, we resonate with what is sexy to them. It's as if, from so widely shared a subjective sense, a strange objectivity emerges. Beauty is both "arbitrary," as we've said, and yet not wholly in the "eye of the beholder." The sense of the beautiful is widely a shared sense. It is quite as if beauty has to it a deep universality. As Darwin wrote, "On the whole, birds appear to be the most aesthetic of all animals, excepting of course man, and they have nearly the same taste for the beautiful as we have."

Why do *we* hear beauty in birds' singings, when the songs are functional only among birds? Why are the decorations and markings that sexually excite hummingbirds and sunbirds so wondrously beautiful — to *us*?

And what of the beauties that plants peddle? For a flower to function, Richard Prum has pointed out, it must in essence join with the nervous system of an animal who will pollinate it, and the animal must perceive the flower as attractive. It is not an accident that flowers appear "beautiful" but roots do not. However, humans are not pollinators, and the messages of flowers don't seem relevant to us. So why do humans find flowers any more attractive than stems or roots?

A human-centered idea of "aesthetics" as a *human* capacity has largely prevented people from understanding that aesthetic capacities, like everything in living nature, evolved.

"Many animals share with humans," notes Prum, "the capacity for aesthetic agency." Their songs are meant to be beautiful, and we too, appropriately, perceive beauty in them. Incisively, Prum observes, "It is not an accident that bird *songs* are generally considered beautiful and that bird alarm calls are not."

Nothing about human sensory systems supports the notion that humans are the sole possessors of the world's aesthetic sensibilities. Indeed, our senses are rather dull compared to the sight, hearing, scenting, and other sensory abilities of many species. Many other animals see *more* intensity, detail, and color than do we.

The world is overflowing with beauties inaccessible with human senses. Humans see light and sense color in the central electromagnetic spectrum, at wave frequencies from 400 to 700 nanometers. Below and above remain invisible — to us. Various other species see above 700, the infrared. Many birds see below 400, ultraviolet light that to us is simply not visible. Their feathers fluoresce with colors unavailable to the human eye and mind. We might see males and females of some species as alike; to them, they might look as different as shirts and blouses. We might see certain birds as extravagantly colored. They might see each other as dazzlingly radiant. They see in each other things we cannot see in them. And that's fair; humans see in each other things that cannot impress birds' sensibilities. But to seers of ultraviolet, flowers and even leaves that to us are simply a wall of green look much more dynamic. Many plants have evolved flowers whose parts absorb or reflect ultraviolet light, forming patterns like "rings, bull's-eyes and starbursts," wrote Ferris Jabr in the *New York*

Times Magazine. Creatures such as we are blind to those particular patterns, but "for many pollinators, they are unmistakable beacons."

Thus, we are not physically equipped to perceive various dimensions of floral, avian, and other beauties. They broadcast extravagances invisible to us, for which we are not the intended audience. Yet when technology reveals those ultraviolet patterns to human eyes for the first time — we gasp.

Deeper now. The case for beauty as a bodily manifestation resulting from sexual choice is convincing, the evidence overwhelming. Yet how are we to explain our attraction to moonlight or a starry night or the holy silence of a cathedral forest? The mating song of a loon is hauntingly beautiful, but so is a meteor shower. We perceive profound beauty in things incapable of trying to get noticed, or of benefiting from attention. Whether the moon inspires us to feel romantic is of no consequence to the moon. Female choice can explain beautifully feathered males, but beautifully feathered males can't explain our response to sunsets.

Like a male animal, a flower trucks entirely in the sexual business of vying to be chosen by a pollinator. But we are not descended from pollinators. *Human* appreciation of the sight and smell of flowers serves no practical purpose, yet we so love the sheer look and scent of flowers that we use them to express our most intensely felt emotions of love and of loss. Perhaps we've somehow fallen prey to the sexuality that is the flowers' stock-in-trade. But we find butterflies beautiful too, and so much else that we have no *use* for. How do we explain and account for all that we find beautiful — and for all that is beautiful that finds us?

More is going on than a misplaced appreciation of sexual lures intended for other species. Snowcapped peaks, blue skies and ocean horizons, shining rivers, a pretty stone —. Far beyond the sexual, beauties in multitudes lie arrayed. To all our senses, the world, the universe, triggers infinities of beauty.

Why?

Chlorophylls, the molecules plants use to absorb the energy from light, absorb reds and blues but reflect greens. Plants look green only because of this accident. Yet for many people, green is the most soothing color. Our bodies and minds resonate peacefully with the main color of the living world. Some people find blue the most soothing: the color of the sky and open water. Research has repeatedly determined that green and blue are "associated with low anxiety levels and the qualities of being comfortable and soothing." These are not rare hues; they are the most abundant colors in the world that made us. And some of the most soothingly beautiful sounds: a running stream, raindrops, ocean waves, a breeze in the treetops — the common soundtrack of the physical world.

I think the evidence leads to a stunning conclusion: the world appears beautiful so that the living may love being alive in it. Life has developed — and we have inherited — a sense of the beautiful to let us feel at home in the world, without further reason.

Beauty is not superficial, or "mere," or a luxury. Beauty is the birthright of living beings. Imagine the unrelieved drudgery of a life without beauty. Subtract beauty, then consider all the grim imperatives and demands of finding food and shelter, competing, procreating; who would want to bother? Emerson wrote, "He thought it happier to be dead / To die for beauty, than live for bread." Beauty is the thing that makes life worth the time it takes. Beauty makes life worth the effort, the risks and the frights and the struggles that being alive requires. Beauty is the reward our brains give us for making the effort to stay in the world. Beauty is what *eases* that effort into joy. Beauty makes our smiles, and gets us past the tears. I think it's that profound, that fundamental. I think that is what all beauties have in common, from the sight of a macaw and the song of a thrush to the deliciousness of good food, the touch of a loved one, or the fidgets of someone small who needs their diaper changed. So maybe we could write, "She found it happier to be here / To walk in beauty, than shrink in fear." Beauty makes us love what it takes to live.

Now let's take the inquiry into beauty all the way to the origins and ask how minds could ever have first developed a capacity for perceiving beauty. After all, there is no beauty without perception thereof, but there can be no perception before there is "beauty" to perceive. Which came first? It's a beautiful mystery. But that's not a very satisfying conclusion.

Wild autumn skies and the sounds of flowing brooks and blowing breezes are beautiful now. But how did living things *begin* to have a sense of the physical world as beautiful?

Something drove the origin of sight and hearing. Something had to drive the beginnings of the perception "beautiful." Some survival value had to exist — an advantage to making the right choice and a cost for the wrong. Any choice requires a surge of motivation (fright, hunger, lust, pleasure). Life does not open and close like an automatic garage door. Garage doors have no reward centers, no dopamines or opioids. But the nervous systems of worms do. The requisite hormones and neurotransmitters for sensing beauty have been in animals for about seven hundred million years. They probably arose, researchers have speculated, "when animals became mobile and started to make experience-based decisions." Even nematode worms have what researchers call "moods."

Hunger is a motivator for getting food. Very basic. No obvious aesthetic there. But animals can perceive food as tasting "good." That is an aesthetic feeling. "Bad"-tasting things might be inedible or spoiled. That's suddenly a basic choice — what to eat — based on an aesthetic. When our dogs are not hungry, they *enjoy* a good-tasting treat. A mind that instructs a body, "Enjoy food, even when not hungry" is less likely to starve. Perhaps the ability to *enjoy* what helps you survive helped drive the evolution of the ability to have preferences, to discriminate. Tasty or bad-tasting, comfortable or uncomfortable, and so on. Once the capacity — aesthetic capacity — was in place, it could eventually be applied in utilitarian ways in preferences for everything from shiny feathers that indicate health to sensible shoes. Or in extravagant arbitrary ways: preferences for impractical outlandish plumes and stiletto heels — "according to [our] standard of beauty," to recall Charles Darwin.

Animals, because they are mobile, must choose the right habitat. They must choose to be in the right place. To make that choice, it must look and feel "good." Macaws stick to forest with plenty of fruit and enormous ancient trees. Humans prefer an open expanse with a view of water (as real estate prices prove). Habitat selection becomes aesthetic: a sense of being *home.* The emotion we call love is a feeling telling us we are "at home" in someone's arms. Think of the simple, profound happiness that comes from being in "the right place." And how *beautiful* that is. Being at home — whether on our native land or in our kitchen — can fell us with fierce swipes of exquisite pleasure. When we look into a dark night sky, the mere sense of our existence, here, among the stars, can completely wash us off our rafts.

As I'd mentioned, there is no natural "selection." Natural "selection" is just filtering; the environment has no likes, no preferences. But there *is* sexual selection, a very active evaluating, coming to a choice. Life has, in the very realest ways, elected to choose arbitrary, random acts of beauty. This is Life itself taking charge of directing evolution along a major path, Life wrestling successfully to take a major role in directing its own destiny.

This is radical. And the radical preference of Life is: beauty. Not only that, the hormones and the bonds show us that many other animals are capacitated not only to experience beauty but also to feel love. Life has created, then consistently moved toward, both love and beauty. Those two capacities are the living world's two Truths, capital *T*. If Life, in its long and difficult journey, has come to be about anything, it is about the progress of love and beauty, the tingle of those Truths.

A taste for the beautiful exists as a deep capacity, one bequeathed to us through inconceivable ages, shared to varying degrees by many creatures. It seems to me that a sense of the beautiful exists to let living beings feel at home, happy and alive, here on Earth. If anything is more miraculous than the existence of life, it is that Life has created for itself a sense of beauty.

And so, behold these macaws, having such fun dangling around. Aren't they, in their splendor and playfulness, just so surpassingly *beautiful*?

I'm sure they think so.

Realm Three:
Achieving Peace
CHIMPANZEES

■ ■ ■ ■

Chimpanzees are always new to me.
— Toshisada Nishida

A great racket breaks out, with terrifyingly guttural roars and thunderous stomping. The chimps show strength to the enemy group by breaking off the biggest branches they can manage, dragging them around, and throwing things in their enemies' direction. Then another round of distant calls sets off an absolutely fever-pitched eruption of hooting and shrieking from the group we're with. From all directions, the air reverberates. Those above scurry out of the trees like firefighters who'd been dozing one moment and in the next are summoned into action, sliding down toward the emergency. We see chimpanzees' contagious excitement and contagious fear as easily as we see the bright side of the moon. The hoots and shrieks are designed to get attention, and because they do, the more subtle side of chimpanzee emotionality gets overshadowed. Yet the deeper nature of chimpanzees includes tender, empathic concern for others and courageous altruism. Always there, seldom noticed.

CHIMPANZEES

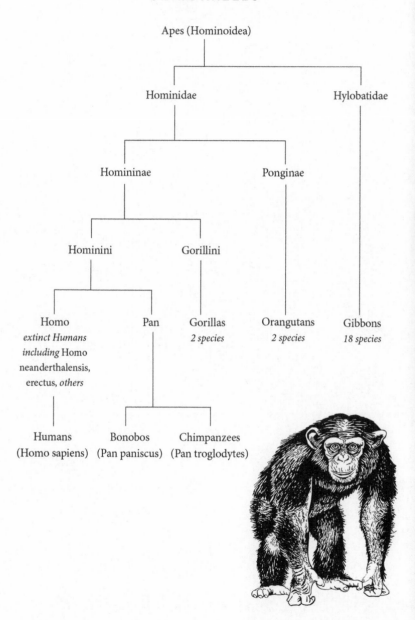

Apes (Hominoidea)

Hominidae

Hylobatidae

Homininae

Ponginae

Hominini

Gorillini

Homo
extinct Humans
including Homo
neanderthalensis,
erectus, *others*

Pan

Gorillas
2 species

Orangutans
2 species

Gibbons
18 species

Humans
(Homo sapiens)

Bonobos
(Pan paniscus)

Chimpanzees
(Pan troglodytes)

PEACE:
ONE

Before I can shoulder my pack, Cat and Kizza vanish into the undergrowth as though swallowed into a green portal. I dart in and scurry toward the sounds ahead. Intermittently I glimpse Cat glancing over her shoulder to make sure she hasn't lost me.

I catch up, and she whispers, "It's Alf. Alf can disappear."

Alf is disappearing now; we must hurry.

He'd just eased his dark, human-like figure down a thick vine, planted his feet on the ground, lowered himself to the knuckles of his hands, then straight and purposefully walked off into the veiling vegetation. Now he's moving at a challenging pace.

Alf's all-fours gait is well suited to these thickets. Upright, on two tippy legs, we're slowed by head-high sprays of leafy branches and spirals of ankle-tangling vines.

Cat and Kizza glide through the thick greenery. I don't have the hang of pushing through, or of untangling wiry vines that seem determined to trip me — and sometimes succeed. But if I lose my companions, I'll be completely lost on my first day in this African forest. Coming back to find me would disrupt their research for the morning. They can't afford that loss of time. I can't afford that kind of a start to my visit. I speed to keep them in sight.

Alf — who is verging on adulthood at age nineteen — is also on the verge of ghosting from us into his world, a world I am struggling to enter.

After a few minutes Alf slows, and as I catch up I get a good view of him. By following Alf, we're studying how chimpanzees manage their day and their social agenda. Cat Hobaiter's life's work is to observe and unpack how chimpanzees here in the Budongo Forest of Uganda use gestures — often very subtle ones — to communicate. Expertly aided by research assistant Kizza Vincent, Cat looks deep into chimpanzee life to discern answers to questions such as: What constitutes a meaning-bearing gesture? Which chimps use them, when, why, and how? Does gesturing help maintain chimpanzee relationships and the social ebb and flow of their days?

Alf joins several other chimpanzees. One, backlit up in a high tree, has a clinging baby.

I raise my binoculars. Kizza whispers, "Dat is Shy."

"How," I ask, "can you tell who that is?"

He replies in light zephyrs of speech: "If I see your shape, you walking away, I know it is you."

It's that easy?

Cat affirms. She just glances: face, build, stride — recognition is instant.

The chimpanzees certainly differ. Some have a boxy body build; some are lanky. Their complexions range: pale, brown, mottled, freckled, coal black. Faces can be flatter or rounder, have heavier or finer features, differing shapes and wrinkle patterns and higher or lower hairlines. Ears, mouths, the shapes and shades of lips — all vary. Hair varies, butch to plush. And that's just here.

Scientists now recognize four regionally differing chimpanzee races across Africa. The eastern, western, central, and Nigeria-Cameroon chimpanzees live very differently across central Africa, in landscapes ranging from heavy forest, like Budongo, to sparsely treed semi-savanna. Humans call them all "chimpanzees," but they have long traveled divergent evolutionary and cultural trajectories. Central and eastern chimps were probably connected one hundred thousand years ago, but western chimpanzees have been separated from them for five hundred thousand years. Most scientists see the four groups as different races, their salient lifestyle differences as mainly cultural. Others think the genetic differences are sufficient for considering them different species. What we can say with confidence is that for a long, long time chimpanzees have been works in progress. "We've learned," writes Craig Stanford, "not to speak of 'The Chimpanzee.'"

Here, in their easy mix of pale and dark faces, chimpanzees have a solution to humanity's fundamental obsession with skin differences. Chimps suffer none of that awful thing; we alone saddle ourselves with those manufactured hatreds. But as I'll learn, they have their own self-inflicted issues. I'll also learn: a preference for peace, within a penchant for war, is another thing we have in common.

Masariki, an adolescent male with distinctive oval eyes and a flat face, rests with his usual companion, Gerald. When Masariki was young, Gerald would help him across gaps in the trees by bending branches or by using his own body as a bridge. Masariki was an orphan. He and Gerald might be siblings, but Gerald is dark-faced and Masariki a lighter shade of pale. Gerald probably adopted him.

Fifteen-year-old Daudi appears with twenty-year-old Macallan, who is missing his left thumb. Cat brings us startlingly close. The guys strut their stuff a bit, and in a moment they're right among us.

I wonder out loud why they are not apprehensive of me.

"Actually," Cat informs me, "they've been watching you. But they don't fear people who are with me and Kizza."

Cat Hobaiter has been working here in Budongo Forest for a decade and a half. Athletic, late thirties, sporting a dark bob of hair, she is, among other things, a very strong walker. A child refugee from the war in Lebanon, Cat eventually earned a doctoral diploma from the University of St. Andrews in Scotland, where she is now a professor. When she first came here, she found chimpanzees, she says, "addictive." Of her motivation she adds, "I haven't come to work in a remote Ugandan forest because I am interested in 'clues to human evolution' or 'how we became toolmakers.' I'm here because I'm interested in chimps." Cat is a highly competent observer because she arrived *without* a lot of preconceptions. "I didn't come here thinking, 'I've read that chimps do this, now let me look for it.' " Because Cat looks for what she has never seen, she sees what others have overlooked.

Chimpanzees seem very familiar to most people, of course. And as with most "very familiar" things, if we stop to think for a moment, we realize we actually know almost nothing about them. We know they are cute as babies. We may think of Jane Goodall lovingly cradling an adorable young chimp (yet not wonder what happened to its mother). We may know that they live normally in equatorial Africa and very abnormally elsewhere, including medical labs where little was learned during decades of tormenting thousands of them to the point of broken insanity. Some people know that they can be "retired" to "sanctuaries," though a retirement sanctuary is merely a kinder captive dead end, and the only real place for a chimpanzee is in a free-living community, an existence essentially impossible for a caged chimp to regain. Some people have heard that certain activists are trying to get chimps declared "legal persons," but most of us are confused about why. (Answer: because a legal person is the only thing that can have legal rights; anything that is not a person can be a person's property.) That accounts for about 99 percent of what 99.9 percent of us know about the creature who shares more than 98 percent of our genes, humanity's closest living relative.

We see in them partially formed prehumans caught between being and becoming, a harbinger of humankind. But to see them that way is to misread history and to fail to see — them. Chimpanzees are not our ancestors;

the last species ancestral to chimpanzees and humans is extinct. Chimps are our contemporaries. They are complete chimpanzees, not half-baked humans.

About six million years ago, our common ancestor hit a fork in the tree, and *Pan,* the genus that includes chimpanzees and the bonobo, and the lineage leading to *Homo,* the human line, began separate trajectories. (Gorillas had begun their journey much earlier, about ten million years ago. Orangutans, roughly fifteen.) Various *Homo* species — possibly up to twenty species of humans — evolved and flourished. They include the widespread *Homo heidelbergensis,* Asia's "Denisovans," *Homo naledi* in South Africa, many others. *Homo erectus* first mastered fire making. Neanderthal bones with healed breaks and evident disabilities indicate that species' care for the disabled. Some of those human kinds disappeared; some became us. Our species, *Homo sapiens* — not gorillas — is the closest living relative of the chimpanzees and the bonobo.

Together with gorillas, chimpanzees, bonobos, and orangutans, humans are in a family called the Hominidae, or the "great apes." Yes, humans happen also to be "great apes." Compared to chimpanzees, human brains have no new parts, and they run using the same neurotransmitters. There is similarity, organic continuity, and overlap. DNA sequencing showed that human and chimpanzee DNA is between 98 and 99 percent similar. Because the genomes contain about three billion nucleotides, the small gap in percentages amounts to tens of millions of little differences. Chimpanzees and bonobos share more than 99 percent of their DNA, yet their social behaviors differ strikingly. Overwhelming genetic similarities reflect close organic kinship. When human sensibilities resonate with a chimpanzee's behavior, it reflects an evolutionary history almost entirely shared, identities largely identical.

"Chimpanzees live long, interesting lives," writes primatologist Craig Stanford. The social world of a chimpanzee is a complex web of friends, relatives, and ambitions.

Cat explains: "It's not just a matter of who you like. It's also: Who are your allies? Who is low-risk for you and for your children? Who knows the food trees — ?"

Chimpanzees often move in groups. Parties are fluid; rules are few. You can be with whom you please. Certain individuals spend part of every day together; others spend part of each day alone. Females with newborns may take a solitarily sojourn.

There *are* also rules. One: mothers and young children are inseparable. Two: though parties are fluid within a community, which *community* you belong to is rigid. Three: male rank matters — a lot.

The basic social unit of chimpanzee life is the "community." A community holds and defends territory — sometimes violently — against other communities. Yet seldom are more than a third of the individuals in a community together at once. Somehow they know exactly to which community they belong. "Sort of blows my mind a little bit," Cat confides.

As with us, culture and group stability depends on a mental *concept* of "we." Crucially, young chimps watch Mom's social interactions, absorbing protocol from her experience, learning how to act deferential or dominant, who they'll be with, where to go, when, and situations to avoid.

Male chimpanzees remain in the community of their birth throughout their lives. Males anchor a community to its land, its customs, and its identity through generations, over centuries. Most females move at adolescence, permanently, into a nearby community. This often means that a female will join up with chimps who have been her birth community's territorial enemies. It's a fraught transition. Upon arrival in her new community, where she will likely live for the rest of her life, she might be welcomed, attacked, or harshly assimilated while suffering bullying by senior females.

Basically, chimpanzees live in tribal groups on tribal lands. Humanlike, not human;

contemporaries, not ancestors. With them we share a deep history sufficient to resonate, to delight, sometimes to horrify.

In this exclusive forest citadel, chimpanzees have maintained an unbroken life and culture for countless thousands of years. Budongo Forest, elevation three thousand feet, is an island of green surrounded by the smoky haze of many impoverished horizons. The rising sea of human settling comes right to its boundary, as though the forest is the drowning shore of a shrinking continent. Budongo still occupies less than two hundred square miles (460 square kilometers), twenty-five miles across the long way. Not much. Though Budongo Forest Reserve is one of the largest forests remaining in Uganda, it was heavily logged for mahogany and other woods, much of which was exported. No elephants remain here, no leopards. People may come and gather medicinal plants and firewood. People may not cut trees or set snares for bushpigs or the small forest antelope called duikers (rhymes with "hikers"). But people do. Ironically, enormous fig trees that grew up in the gaps of the twentieth century's intensive logging now provide more food for chimpanzees at certain times of year than did some of the native trees that were logged off. This is helping the chimps persist with less space.

The researchers here have named several

chimpanzee communities after parts of the forest. This community that's the focus of Cat's research is Waibira. Adjacent to Waibira is another well-observed community, called Sonso. Sonso, the smallest chimpanzee territory known anywhere, is only about three square miles (eight square kilometers). Waibira, at twice that, is still significantly smaller than average. Elsewhere, most chimp communities occupy territories of around eight to ten square miles (twenty to twenty-five square kilometers). The seasonal fig abundance notwithstanding, food in Budongo is a concern. Over twenty years, the amount of fruit has declined by an estimated 10 percent, seemingly due to climate warming.

Waibira is made up of about 130 members, including an unusually high proportion of adult males. Sonso has about 65 individuals. Sonso's shared border with Waibira sometimes puts them at odds, even though it also means each community contains some females born in (or destined for) the other community. Life is complicated. Sonso also borders the forest's edge and farmers' fields, creating two-way problems. When Cat first came here, the forest was different. People have pushed farms into the edges of the wild land, of course. And some illegal logging continues. Chimpanzees relied on different kinds of trees for fruits and seedpods that ripened at different times of the year. But certain trees highly valued by chimpanzees are also worth money, and have been drastically depleted.

Our research station is a cluster of ground-level dormitories on the site of the former sawmill that dismantled much of the original forest. Varied birds and animals are accustomed to the camp and its humans, lending the place a Garden of Eden air. It's normal to open your door and have two monkeys streak past your toes. Baboons of olive hue, blue monkeys with nervous eyebrows, red-tailed guenon monkeys whose striking facial markings look painted on, and velvety colobus monkeys with elegant white capes — all visit daily. Local women bend over wood fires to cook our supper of rice or cassava with beans or peas. We drink rainwater purified through porcelain filters. There is no plumbing. The forest air blends the perfumed and the putrid into a perfect mosaic of mold and metaphor.

The chimps discriminate very well between the researchers and the villagers in their forest. They fear the villagers — with abundant reason — but have become accustomed to the researchers. Cat invested years to simply get the Waibira chimps to ignore her presence so she could observe them behaving naturally. Researchers provide no food. There is close proximity but no interaction. We're just here, and they're used to that.

Eight or nine other chimpanzee communities here in Budongo Forest remain untamed, unnamed, untallied. Vaguely referred to with names like the Northerners, their identities, deep and absolute, are known only by them.

To truly comprehend any creature — including people — you must watch them live on their own terms. Budongo chimpanzees are in charge of their movements and decisions. Their lives are complex, not neatly categorical. There's a limit to what can be learned from captive groups whose lives lack the foraging, hunting, and intergroup dynamics that are the main modus of chimpanzee society. Budongo's free-living chimpanzees are homeschooled in their own traditions, the knowledge of mothers conveyed to children through ages of deep, wild history.

PEACE:
TWO

Ben, the mottle-faced emperor of Waibira, arrives flinging things and dragging dead branches noisily through dry leaves. He — especially in his opinion — is not to be missed. As "alpha," Ben has held "office" since last year. In only his late twenties, Ben was unusually young when he attempted leapfrogging several contenders. "I wasn't expecting him to succeed," Cat offers. "But he made it, early."

I'm interested in a seeming paradox: How does chimpanzee culture facilitate a shared group identity and help the chimps hold their community intact *despite* the internal pressures of intense male ambitions? You'd think members might simply walk away from the arena of threat and potential violence. Their high-strung lifestyle includes the continual drama of males plotting with strategic allies to seize rank and maintain dominance. Such a pressure-cooker-waiting-to-explode system would seem to sow the seeds of its own disintegration by the mere attrition of peaceful parties. Yet something maintains the system and the group's cohesion. As with humans in less than ideal social conditions, there must be some advantage to staying in the group. I'd especially like to understand chimpanzees' cultural mechanisms for disarming and de-escalating conflict, resolving or at least defusing tensions, and usually maintaining a tenuous but tolerable peace.

Ben is shaking saplings, hooting, and thump-kicking the buttress roots of big trees. The high-ranking are seldom secure. They assert and reassert with bluster and noise, because strength is mainly what they've got. But there's a lot of competing strength among chimpanzees. And willingness to use it. And ambition. So, in addition, there's strategy.

"Males who rely only on strength," Cat tells me, "who make a habit of attacking over minor things or taking fights further than necessary — they don't get far or don't last long. Most use a balance; a few are amazing strategists." You can play the game in different ways, she explains. For instance, Zefa remained second in dominance through two alphas. "You can remain in the number two spot and get the benefits, while avoiding a lot of the stress of grabbing and holding on to the top slot. It's worked for him."

Ben's rank requires all others to acknowledge his superiority with a specific greeting called a "pant-grunt." It's required when a high-ranker must be greeted with a respectful "Hello, sir."

But today Alf doesn't pant-grunt to Ben. More surprising: he gets away with it.

"Ben seems to have lost some status, yet he is maintaining rank." Cat appraises their politics with the eye of a seasoned analyst. "He's not getting due respect. He might have just decided not to press a contest he could lose. On the other hand, while Alf's behavior was a slight, it was not a challenge."

Hierarchy is *the* preoccupation of male chimpanzee life. For them as for us, status seeking is an impulse, dominance its own reward. In a fight, allies back each other. If the involvement of allies ends the fight, it's rarely because they are peacemakers; it's because their side has won.

Rising up the male rank order entails calculated risk. Maybe you've been allies with a particular high-ranker for five years; you might switch allegiances for an opportunity to move past him. If that's your decision, you might have your life threatened by someone you've been close with for years.

"There's always a lot of subtle politics," Cat explains. "Who is seen with whom, who sits where, who gets up and follows —. It's like who gets to go to lunch with whom." These little things clue a seasoned observer like Cat to building tensions, well before sudden violence brings a power shift.

A possible reason for Ben's reluctance to assert to Alf: Ben must be unwell. There's been a serious coughing cold going around. Almost every chimp's been sick; a couple have disappeared. Maybe today no one's up for a contest of the hierarchy.

From the deep brush comes a chimp with a black face and a torn ear. This is Lotty, one of the regulars. In her early thirties, Lotty has a six-year-old daughter. Lotty dutifully pant-grunts her acknowledgment of Ben's supreme rank, then begins grooming him.

Lotty and Ben seem absorbed. Until —. They stop. Listen attentively. Are we hearing friends or an adjacent community? They're monitoring who is where, what they're up to, what they might be up against. They always want to know: What's going on in the community?

They move. We follow. They travel on the soles of their feet like us, and on the knuckles of their hands, unlike us. They're moving off-trail, necessitating bushwhacking by us uprights. We walk single file, me last, following Cat.

A towering fig tree bursts above its neighbors, spreading its canopy over them. High in chimpanzee heaven, dark shapes reach black arms across the baby-blue morning.

Because the tree's massive, buttressed trunk is too huge to ascend, Alf climbs one of the small surrounding trees as easily as we ascend stairs, basically walking straight up the trunk. When chimps are climbing, their thumblike big toes give them, effectively, four hands. Their short, thick legs propel them straight up trunks. Their long arms make a ballet of it all. Now Alf spans himself into the fig's spreading branches, where he continues upward toward the fruited heights. Shy begins her own ascent; with a baby clinging to her belly, her strength is even more impressive.

Shy stops climbing and extends an arm. It is her gesture of intent to her baby to use Mom's arm as a ramp onto the main trunk. He grasps a tiny vine and pulls it, holding its tip in his mouth while consolidating his grip on a more substantial portion, clearly understanding exactly what's needed for safety. Hanging by one arm, he swings on the vine, building momentum before making his next reach to the branch he's targeting. This little six-month-old fluffball immediately proves to be a crazy climber, swinging hand over hand through the tree with his feet dangling. He's clearly loving it. His mother watches carefully but seems confident in her child.

Aloft, the chimpanzees are Olympian. They swing and float. Just to be here watching feels otherworldly. Them climbing. The females with babies. Their noisy rustling as they navigate the canopy. We listen to them eating. There's sufficient daylight now to see that the chimps share this tree with half a dozen monkeys and eight black-and-white-casqued hornbills, who caw through the canopy on big swooshy wings. This rooted, towering being not only grows figs; it grows these animals and sets them in motion among its branches. They shriek, whistle, hoot —. If you think trees do not talk, you are partly right. The trees let the creatures they grow do the talking.

The chimps' climbing and swinging, their reaching, picking, and eating, would seem absolutely to require four-limbed proficiency. But I'm noticing that the chimps now aloft include several whose hands or limbs seem impaired. Cat says that about three of every four adult chimpanzees here bear a snare's disfigurement. For many, it's just a scar or a stiff hand. Others lack fingers or toes. Or worse. Thirty-year-old Jinja has lost her right hand; ten-year-old Andrua lacks his left; Philipo is missing a foot.

When a snare's noose closes around a chimpanzee's hand or foot, they immediately pull with all their immense strength. This force sinks the cable (bicycle brake cables are commonly used) though the skin. The panicked chimpanzee screams and spins around and around, trying to get free. Eventually the cable twists to the breaking point, releasing the now-maimed chimp, whose troubles have just begun. Some die of infection inside a week. Some survive much worse. Tatu, whose name means "three," is navigating the branches carefully. Her left leg is gone below the knee.

"It's heartbreaking," Cat says, "to see them using their communication gestures with hands that are missing fingers." You'd think a missing hand or foot would be a death sentence. But then you see one with such an impairment, see that they persevere, climbing despite their handicaps, sometimes with a baby on their back, getting on with life despite us. Cat comments, "Us at our worst; them at their best."

Alf indulges in half an hour of eating high up in the sunshine.

Back home on Long Island, New York, my writing studio is on the second floor of a little cottage. When my dogs are on the ground floor, Chula (but not Jude) will often come upstairs and start banging on me with her paw — as just happened while I was typing the preceding paragraph (hence the abrupt change of subject). When I look at her and say, "What?" she runs to the top of the stairs and looks at me. If I don't get up, she returns and bangs some more. If I get up, she runs downstairs. By the time I reach the bottom of the steps, Chula and Jude are already at the door, facing it with tail-flags waving. Chula's plan for getting *out* the door involves coming *upstairs,* as far *away* from the door as is possible in the cottage. She knows she needs my help to accomplish her goal. So she comes *up* to get me to come *down* and open the door for them. It's quite a simple plan. But quite simply — it's a plan. And that gets us back to chimpanzees, and the complexity of their planning.

Compared to a doggie at a door, life for free-living chimpanzees is vastly more complicated. For example, chimpanzees know and keep track of dozens of food trees and their stage of ripeness. Chimps don't *wander* in search of food. They commute. They head to trees they have in mind. If they check a tree and the fruit is far from being ripe, they'll adjust the timing of their next visit according to whether the subsequent weeks are sunny or rainy, weather that speeds or slows ripening. "It's insane," Cat admits. "I couldn't keep track of all their trees. I couldn't do it."

Alf takes a break from gluttony, lying on a broad limb between two heavy clumps of figs, a simian Bacchus. Life is good for him at the moment. On the forest floor, the filtered light is soft. We rest against trees, watching chimps and listening to doves sending beautifully rhythmic coos to one another.

When Alf ambles easily off across the shadowed forest floor, we again struggle to keep pace through the dense vegetation. Walking against the uneven ground requires a continual looseness, a constant catching of one's balance, like walking on the deck of a rocking boat. We are usually in a hurry, sometimes in a great hurry. The vegetation is in places so thick, I have to part it with both hands, as if swimming. Walking upright remains a distinct disadvantage. The densest vine tangles force us down to hands and knees. Cat has a pair of clippers, and we snip our way through the creepers and draping lianas in cover so heavy it's like being deep in green seawater.

Suddenly we nearly stumble upon Alf. Resting just fifteen feet away, he is half-hidden by dense foliage. Though at rest, he remains attentive to the distant voices of his fellows.

As Cat pulls out her coffee, Alf moves. "Most effective way to get a chimp to travel," she says with a roll of her eyes as she slips the untouched java back into her pack, "is to think you have a moment to relax."

In just a few paces, Alf stops again. He has met up with four resting chimps. But when a round of sudden *wah*s and hoots ignites them, we realize that more like fifteen chimpanzees are hidden in the leafy shadows nearby.

Alf utters a "resting hoo." He makes several long arm scratches to invite Gerald to a grooming session. The sound of his fingernails in his hair makes the gesture surprisingly audible.

Alf sits with one arm out, a gesture suggesting that this would be the ideal spot to commence grooming. Gerald comprehends the request and complies. Gerald jabs four fingertips into Alf's hair and lifts to expose and skillfully remove impurities and any bugs. The jab-and-lift technique is true Budongo style, a cultural quirk here that is unlike the long, raking motions chimps elsewhere use when grooming. After a minute Gerald gives Alf's head a gentle nudge with his hand, like a hair stylist might: "Move your head so I can get to your neck." For twenty minutes, they indulge.

Grooming's superficial effect is parasite removal. But the much deeper function is promoting trust and alliances. The power of touch.

Because touch is potentially dangerous and, therefore, can be frightening, it's good to have an excuse. Remember putting your hands on a dance partner for the first time? Dancing was the excuse, but touch was the reason. Humans have retained some social grooming: combing and brushing another's hair, spreading sunblock; you do these things with people you have some relationship with. It helps deepen or maintain the relationship. You don't do these things with a passing stranger. You don't have a relationship until you invest time in it, and, Cat notes, "If you're a chimpanzee, you invest time by grooming."

Chimps are selective about whom they groom. They often select: males. Males groom males six times more often than females groom females. Females groom males more than they groom other females. When we say metaphorically that a man is "being groomed" for a higher rank, we're intuiting that grooming can be about male power. With chimpanzees — it is. Chimps are said to groom "up the hierarchy." Subordinate males like to groom higher-ranked males. An alpha male gets groomed the most.

Grooming among male chimpanzees is about establishing and maintaining the social bonds that will later be needed for co-operation in territorial defense, hunting, and taking and keeping rank — ultimately meaning preferential access to food and sex. If two male chimps have been allies but one decides to dominate, their grooming relationship will fracture.

In chimps, these concerns are male things. In humans, they're often male things. But in some other animals — macaque monkeys, for example — females do more grooming than males. Human females seem to groom each other more than they groom males and much more than human males groom males. In other apes, either these power relationships don't apply or they're not male-oriented. In bonobo society, power is female; and female bonobos use their power to maintain peace. Bonobos, elephants, and orca whales, for example, automatically confer status to *females,* as they simply age into high rank through seniority. (If you're not familiar with bonobos: they are superficially similar to chimpanzees but are a separate species. Their ancestors split around two million years ago. Socially, bonobos differ fundamentally from chimpanzees, with females dominant over males and with far less frequent and less serious violence. Bonobos live only south of the Congo River. Their range, far smaller, does not overlap with chimpanzees'.)

If the science of watching animals has one great message for humankind, it's that female power — whether in bonobos, elephants, sperm and orca whales, or lemurs — tends to create space for peace. Chimpanzees have actually gotten that memo, but not enough of them have read it.

With the grooming accomplished, Alf walks a few paces and makes long, loud arm scratches while looking over his shoulder at Gerald. Alf's gesture is a request: "Let's travel, together."

Cat says, "They're probably going for a drink." Cat isn't always conscious of the nonverbal cues she's responding to. "Sometimes," she explains, "I visit a different chimp community and suddenly realize that what I expect to happen next, doesn't happen. Each community does things a bit differently."

A while later, we've followed several chimps to the water holes, which in this dry season are a series of mere seeps in a ravine. More continue to arrive in small parties.

Through shafts of light and shadow, out from the undergrowth two or three at a time, come a dozen chimpanzees. Masariki follows with his friend and protector, Gerald. Cat knows each individual at a glance. Monika, Fiddich, Lafroig —. (Cat has named some of the chimpanzees more or less after Scotch whiskeys, taking some liberties with the spellings. The Scots in camp insist that one must never put ice in Scotch. We have no ice. We do have Scotch. So no problem.) Lafroig, an adolescent with big brownish ears, is carrying a fruit-laden frond in his teeth. His unusual food-toting habit has earned him the nickname Chubby.

Some days, they drink and run. Today they've decided to lounge peacefully. For almost three hours they drink, rest, and socialize. I soak in the forest's peacefulness and finally close my eyes.

A sudden commotion startles me to attention: the arrival of Talisker. He rattles vines and drags his knuckles loudly through the dry leaves. By his performance he displaces a couple of females who'd been sitting peacefully. "I don't think he's done anything Ben would object to," says Cat.

Talisker holds an odd position of high-ranking statesman outside the power struggle. Cat describes him as "a silver fox, a distinguished older gentleman who does very well for himself." He and the alpha Ben are not strategic allies. But Talisker is a master strategist of his own fate. Talisker, in his mid-forties, has nearly two decades on Ben; he might have been the alpha male before Cat started her research project. Many deposed alphas plummet in status as they hurtle toward an early death. Talisker survives with a kind of grace and dignity in a category of his own. Having stayed central yet removed himself from the squabbles over rank, he has assumed a kind of emeritus status, a comfortable retirement in which he remains much respected but not contested. And he contests not. He's big, too. Cat considers him "the best-looking chimp in Waibira," and that is saying a lot for a senior citizen who no longer shows the buff muscle of prime-age males. His longish hair gives his shoulders a sort of epaulet look. He maintains his status without competing, and carries himself as if he doesn't worry.

Just as Talisker has begun enjoying a drink with a wadded-up leaf sponge, two young-adult males nervously arrive to pay their respects. Because it's the dry season, the water holes are depleted, mostly mud. Talisker is sitting at a tiny, clear puddle, drinking his fill.

The junior males pant-grunt their acknowledgment of his superiority.

Talisker gestures them away with flicks of his wrist. He could push them out of the whole area, but he's a diplomat with discipline and restraint.

When Ketie walks past Talisker, her pant-grunt of respect is so anxious it verges on a scream. Her nervousness puts sixteen-year-old Fiddich on edge. Fiddich approaches making high screams of submission and extends his arm with the hand vertical, like a human about to shake hands.

Talisker quells Fiddich's fear with an outstretched arm, fingers extended. A confident gesture. All this emotional resonance reflects their ability to read moods — part of the empathy spectrum — and respond with due flexibility.

Talisker is an unusual chimpanzee. But he's not a saint. When Rita, who is about thirty years old, approaches the puddle, she pant-grunts to acknowledge His Highness. He moves toward her — a slight escalation. Fear-smiling with lips back and teeth together, she emits a brief, soft scream. Her thirst seems urgent. So she reaches out with her right arm and presents the back of her wrist, the least vulnerable part of her hand. She's recognized his status; that's all he really wants. But he doesn't move. Then Talisker shoos her off with an emphatic flick of his imperial wrist.

Cat interprets: "She's saying, 'I respect you, I respect you; but I need a drink right now.' There's enough water for them both. But he's being a bit of a jerk about it."

Unquenched, Rita goes to Alf. He's also monopolizing a good little drinking spot. Rita pays her respects. Alf leans back to make room. Rita finally gets herself a long, deep drink. Both Alf and Talisker clearly understand what Rita wants, but Alf is being more of a friend about it. Rita is about a decade older than Alf, but she's fourteen years younger than Talisker. Perhaps a little seniority helps her with Alf. But perhaps it's just personality; Alf is a rather mellow fellow.

When N'eve brings two-year-old Nimba to Alf, Alf greets the baby with an open-mouthed kiss on the head.

We follow Talisker into the forest. Along the trail is a surprisingly relaxed-looking Ben — reigning emperor of Waibira — reclining, propped on an elbow. Talisker sits down

403

about twenty paces away, as if for no particular reason, as if he simply feels like sitting there.

Ben begins making exaggerated scratching motions along his arm. Talisker does exactly the same. Each scratches, looks at the other, and waits for a response. They are each trying to convince the other to be the one who moves closer and starts grooming. The one who moves will show that he accepts and acknowledges that he is lower-ranking.

In his emeritus, elder-statesman position, Talisker does not challenge Ben, or try to displace him. But he seldom acknowledges Ben's superior rank.

Ben gives a loud scratch. Talisker gives a scratch, shifts position, and yawns. Cat calls these ego contests "scratch-offs." Talisker has given the "avoid" look. Exaggerated nonchalance. She explains, "They'll quite deliberately look every which way except *at* the other male."

Ben moves about half the distance to Talisker and sits down hard — *bang!* — in his line of sight, making his point clearly and insistently. There is now no way Talisker can feign not getting Ben's message. This forces Talisker's hand. If Talisker doesn't *clearly* respond, Cat says, "he'll undeniably be ignoring the alpha male."

Talisker seems satisfied by Ben's having moved nearer to him, so he gets up and closes the remaining gap. And, though it would be more "proper" for the lower-ranking Talisker to begin grooming first, they make contact with each other in the same instant. Both faces have been saved. Thus begins their grooming bout, as if they're the best of friends.

Macallan, an up-and-coming young male who poses no competition to Ben — not yet — approaches. So we have three generations: Talisker is in his mid-forties, Ben nearing thirty, Macallan just twenty. Macallan tries to initiate grooming and play with Ben. But Ben doesn't reciprocate. Perhaps he sees a potential rival.

And what's on Macallan's mind? "I don't think," Cat considers, "that there's a conscious strategy like, 'Okay; I'm twenty; what do I need to do to get on track for being alpha someday?' "

Occasionally, though, there comes along an amazing strategist. Zefa was the second-ranking male — the closest ally — of the previous two alphas in Sonso. By being steadfastly number two, he never had to assert himself (an alpha is constantly on guard for a challenger). But by making himself indispensable as an ally, he got "alpha level" access to, as Cat puts it, "meat and girls."

During the alpha transition from Duane to Nick, Zefa — who had previously had nothing much to do with Nick — switched his al-

405

legiance on a dime. He and Nick became inseparable. And it all paid off. Zefa became one of the most successful fathers in Budongo.

The first inkling Cat had that Nick's tenure was getting a little rocky was when she saw Zefa buddying up to the most likely challenger, Musa. It was subtle. Zefa would choose to sit next to Musa, and they'd groom — but only when Nick wasn't around. Cat once saw them jump apart when they heard Nick coming, then sit in their new positions, carefully ignoring each other. Cat says it was like a husband getting caught flirting. "I mean," she recalls, "Zefa genuinely sprang apart, like, 'Nothing happening here.' It's still the closest thing to showing a sense of guilt that I've ever seen in chimps."

Musa and Hawa were the two most likely contenders to succeed Nick. Musa is a big, no-nonsense male who carries an air of authority. Musa *looks* like born royalty. Zefa logically backed Musa. Nick lost his alpha status a few months later. But instead of a clean coup, the takeover became a running struggle between two fairly evenly matched males, Musa and Hawa. Hawa won and became alpha — surprising Zefa and everyone else.

For the first time in about twenty years, Zefa had miscalculated. The cost was immediate. He dropped rapidly down the ranks, having to pant-grunt to young males he wouldn't have even bothered to acknowledge a few years earlier. Zefa had followed a clear and sensible strategy. But he'd bet on the wrong horse in an upset race.

Having been ignored by Ben the Supreme, Macallan turns to Talisker and begins grooming him. Grooming can get you close to someone higher-ranking. This seems Macallan's immediate goal. In a few minutes, Ben and Macallan, facing away from each other, engage in a couple of minutes of idle footsie. Then Ben reaches out his hand for Macallan's foot and tickles it. Mission accomplished by Macallan. If Ben sees in Macallan a potential rival, he's keeping him close and on good terms.

If you're lower-ranking and you approach, recognize, and greet the arriving male, he'll usually leave you alone. But it depends on personality. Different alphas set different tones. Cat elaborates: "You get all the classic archetypes. You get the big, strong male. You get the tactical politician." Duane mixed those aspects, and he remained alpha for seven or eight years — an unusually long tenure. Average tenure for an alpha is four years. The record: sixteen years for the male Ntologi in Tanzania's Mahale area.

"I guess 'leadership styles' is the best way to say it," Cat offers. "Their career depends,"

407

she says, "not just on what they are. It depends on *how* they are." Some top males interact mostly with males. Some are more egalitarian. Some are very domineering.

Good leaders keep their top spots longer. And when they fall, they don't fall as hard or as far. Of the former Sonso alpha Nick, Cat says with brutal frankness, "He was not a very good chimpanzee." He'd create disputes and, she adds, "push conflict beyond what was reasonable" (implying that most chimps are, on balance, reasonable). Nick would ignore those who wouldn't acknowledge him — but he'd challenge chimps who'd shown due deference. If a younger chimp acted deferential and then scooted away in fear, Nick would give chase and attack them as they were running away. Many a male in Sonso bears Nick-inflicted scars on his back.

One August day, Juliet appeared with her newly born baby son. She'd been away for some months. Juliet pant-hooted at a distance of forty yards. Nick puffed up aggressively, forcing her up a tree. He followed her from tree to tree. Her screams brought Zefa and Hawa to protect her from Nick's attack. Despite their attempts, Nick continued attacking her over the next half hour, beating, kicking, and biting her. Nora joined the attack. It was she who actually killed Juliet's baby. Later, Hawa picked up the dead baby, placed it on his own belly, looked at its face, and groomed it tenderly.

Two years later, the scene repeated, with Nick relentlessly chasing Juliet and her new small baby from tree to tree. "When she couldn't run anymore," Cat relates, "she approached Nick, pant-grunting." Nick instead grabbed her, hitting her and biting her on the hands, legs, and back. Nambi interfered but failed to stop Nick. Other females, all screaming, were inclined to intervene — but they feared Nick. Hawa chased Nick away, then groomed Juliet for half an hour. Some chimpanzees become troublemakers; some are born peacemakers.

A field assistant who watched Nick grow up says that Nick was always getting beaten up, and that because he was bullied he became a bully. The form of dominance he grew up with, he applied. Even in chimpanzees, apparently, abuse can perpetuate abuse and lead to a kind of toxic masculinity.

"Nick was a terrible, terrible alpha," Cat reiterates. During border patrols, upon contact with the enemy he'd run away screaming. "On the front lines he'd leave females with babies clinging to them while he, the alpha male, took off." But females like protective alpha males, the kind of males who keep harmony and tend to enforce peace. No one likes a bully. So Nick didn't hold the alpha position for long. Often when a chimp loses his alpha status, his position in

the hierarchy drops to second or third. Nick's
plummeted. "Straight down," Cat recalls.
"And everybody made sure he knew it." He
soon died.

PEACE:
THREE

Nothing can come to light except through darkness. Close your eyes. See?

A baboon barks me out of my dreams. Before I sense any hint of daybreak, the tree hyraxes begin another round of calls. Their hoarse screams in the predawn gloaming are my real alarm clock. My phone's alarm is never what actually wakes me; it merely reminds me to put my feet on the floor and get dressed and moving.

I leave my little house and walk through the shrieking darkness for two minutes, to the pantry that doubles as our eating room.

Cat and I quietly sip bad instant coffee in the dimly lit room while we await the arrival of Kizza and another research assistant. The research assistants live outside the forest, in the dusty, hardscrabble settlements among the sugarcane and the maize stubble, about half an hour's bicycle ride down the dirt road where bent women trudge beneath backloads of firewood or somehow balance impossibly heavy jugs of water on their steady heads.

I break two of the small, sweet local bananas out of the screened cabinet, inspect them for rodent damage, and slip them into my pack. In recent days, the avocados have met with hearty mouse approval; luckily for the bananas, the avocados are in a different cabinet and the mice have focused their attentions there.

In fact, when I came in this morning, there were two mice trying to get *out* of the other cabinet. I obliged them, which was good for both them and us. I guess someone closed the doors last night while they were already picnicking inside.

While we wait for the field assistants to arrive, I wander out and look skyward. The moon, nearing full in this dark-sky place, has been bright all night. Shortly before dawn, it has set, giving the peculiar effect of night getting distinctly darker even as the eastern sky lightens. Today it really is darkest before the dawn. Stars briefly crowd our thin wedge of eternity while I gaze up from the surface of planet Earth. Beyond our planet lie only facts; all meaning resides here.

Cat comes out, making sure to close, latch, and double-check the door. In half an hour, baboons will start systematically testing every door in the compound. They'd love nothing better than to get into the room with our food stores and leave us an epic mess. Even though it's been a year since the last security lapse, they obviously remember what riches and fun await behind a slide bolt forgotten or a lock unclicked. Young ones, of course, learn by following the adults' lead. It would be unimaginably delicious for any young baboon to gain access to a cache of candy or food or to ransack any pile of clothing. "For the little ones, every aspect of breaching human security seems like fun," Cat says. "They don't act like it's work to try all the doors."

Indeed, the baboons daily try to pull the mesh off our windows. Humans provide new kinds of habitat and new opportunities for innovation. Nothing in the wild would prepare them to drink by dipping their tails into our rainwater cisterns, then licking them, but they've figured it out. Humans speed the rate of change, sometimes to the breaking point, but these baboons have kept pace with us. They patrol our camp constantly, at close quarters, no more afraid of humans than any skittish dog would be. They're cheeky and calculating, always with a plot up their metaphorical sleeves. This baboon group's culture now includes a collective relish of larceny.

Having secured our rooms and pantry stores against simian disaster, the four of us shoulder our daypacks, walk to the edge of the towering, pitch-dark forest, flick on our headlamps, and ease our way in. The day starts (and will end) with a three-mile walk into the forest.

Much of our morning walk entails a long, slow uphill plod in the dark. As the sky overhead lightens just enough to quench the dimmer stars, a low rumble rises. Not loud. It arrives from everywhere and nowhere in the forest. It sounds oddly familiar, but —. I look at Cat. "Colobus monkeys," she explains. Ah. And now I realize what's familiar. Their roaring sounds like a low-volume version of another monkey from half a world away: howler monkeys of New World tropics, whose gravel-in-a-blender roars ushered in our days among macaws in the Peruvian Amazon.

The exertion of our inclined trek has me breaking a sweat, feeling my shirt back getting wet while my fingers remain chilly. In the coolness I alternate between overheated and chilled, then feel overheated *and* chilled simultaneously. When we're able to switch off our lamps, I can see my breath.

A forest at dawn can seem like the quietest place on Earth. Yet it sparkles with sound. While the infinity of stars above underscore vast vacuums of space, the notes of Earth's living come across the quietude as frogs and trilling insects yield to birdsong. Susurrations and declarations, greetings and exclamations, all the living lit in conversation. Silence is not the absence of sound. It's the absence of noise. There are reasons to love so magic an interlude. But residing deeper than reason is the felt music of such plush silence. Dawn is the song that silence sings. In a recess of the world such as this forest, you can still hear the magic. Outside such whittled hideaways, one species fills up all of the spaces between the notes. Nonetheless I am cheered by the thought that as the eyelash of daybreak rolls endlessly across the planet, a chorus of birds and monkeys is eternally greeting a new dawn.

High in the branches, it is morning. But down here at the forest's seafloor, night lingers. Even as the first chimp trapezes through the canopy, a bat flutters past my face. As the sun climbs the angles of morning, light filters down into the forest like sunbeams dissolving in deep sea. While that tropic sun begins to sear the canopy, we continue to walk in shadows on the deep forest floor as though bathing in cool pools.

Under the powder-blue sky of a world just opening one eye, we press on to the water holes. Cat and Kizza have set hidden automatic cameras around the main drinking spots to capture portraits of the most peripheral females. Surprises await.

Cat's automatic cameras occasionally record images of three females she has never seen. Some chimps, such as Lotty, move with a social core group we encounter daily. Others, though retaining membership within this community, stay aloof for months, sometimes years at a time. Preferring the company of their children and one or two familiar friends, they seem to be introverts, or perhaps they just abhor the males' obsession with hierarchy and the histrionics of life downtown. It's quite as though some of these chimps, in their more perfect freedom, are masters in their own individually devised art of living. "Gladys hadn't been seen for a few years," Cat says. "Then she rocked up with a two-year-old baby. We see Virunga about every four years, when she's in estrus. She comes, gets pregnant, and disappears for a few years."

Pregnancy lasts eight months. Because the infant's head is small relative to a human baby's, and the chimpanzee pelvis is not specialized for upright walking, birth is merely uncomfortable, not difficult. Into her own waiting hands, the mother delivers. She bites the umbilicus, then expels the placenta. She either eats the placenta — sometimes sharing it — or discards it under some leaves. The newborn, helpless as a human, will for the first couple of months be held inseparably to the mother's body. Exploration, play, and socialization — carefully monitored by Mama — will begin at around three months. The young nurse for about five years, stay with their mothers for their first decade, become independent of their mother when they can match the pace of adult travel, at about ten years old, and begin to act fully independently at around fifteen years. Females tend to give birth every five years. Female chimpanzees do not experience menopause; they can give birth into their forties and fifties, their limit of lifespan.

Females about to give birth or with newborn infants sometimes make themselves scarce during this delicate time, while their tiny newborn would be at risk from the overzealous interests of others.

"To chimps, a new baby is the *most* special thing," Cat says with a pleased smile. "Any new baby sparks great interest in the chimp community." When Cat first came, births were infrequent. "All the chimps were so excited by every baby. And for us it was incredible cause for celebration."

And?

"Then the bad stuff came."

Sonso has had several infanticides; Waibira has had one. Sometimes there's no infanticide for ten years — then suddenly several. The rare — though unpredictable — darker infanticidal impulses of (usually) males might explain why many females go on "maternity leave." A baby less than a week old whose mother has been living peripherally or who has just returned from a long sojourn is most at risk. Sheer overenthusiasm creates added risk. Everyone wants to touch and handle the baby. Some chimps are too vigorous in greeting an infant. They can get carried away with their aroused interest, pulling and poking. Accidents can happen. But sometimes a male swings a baby's head into a tree with intent to kill.

In some other mammals, such as lions and bears and even some squirrels, a strange new male sometimes shows up and kills the resident infants. (In wolves, however, new males *adopt* existing young ones.) What makes chimps different is that a male who has known a female for twenty years might kill her baby. Violence from *inside* the community; that's what's unusual about chimps — and about us. Humans are the other large primate capable of murder, infanticide, and violence from within a community.

In chimpanzees, males coming up the rank hierarchy seem to account for most infanticide. Whether it's just impulsiveness or displaced frustration, no one knows. Luckily, infanticide is rare. But Cat knows firsthand that when it happens, it's so upsetting it can stick with you for a long time.

"Nowadays when a new baby is born," she says, "my first response is dread: Will they be all right? When they get to be six or eight months old and you see them playing with the older males, you can exhale. You think, 'Okay, we're good. We can give them a name.'"

■ ■ ■ ■

On this new morning, Ben and Talisker —
whom Cat refers to as "the big boys" — are
not with the first group we've found. And
now Macallan, age twenty, performs a bit of
male bluster, dragging stuff around. But he's
almost entirely silent about it. He wants to
impress onlookers without summoning Ben's
rage.

Macallan's show frightens year-younger
Sam, who runs partway up a tree.

But Lafroig, *five* years younger, begins
displaying too, and so Macallan and Lafroig
are displaying *around* each other. Lafroig is
being very careful to make a show of himself
while avoiding affronting Macallan directly.

This is the slow, years-long boil of chimp
ambitions.

These performances agitate the females,
who'd been restfully eating. When Lafroig
displays *at* one of the females — she screams.

Her scream brings Ben suddenly barreling
through to make everyone knows who's boss.
Soon Ben and Macallan are streaking through
the trees and then racing across the ground
at top speed. Ben means it, and when they
disappear into the undergrowth, Macallan's
wails are audible for at least a hundred yards.

Large males who were playing peacefully
with babies abruptly chase off the now-

screaming little ones. Mothers begin defending and collecting their kids. Everyone is hooting, screaming, running up tree trunks. It's chaos.

And what started it? Macallan had simply displayed.

It's easy to say, "That's all it was." But to them, it's a big deal. To the males, it's their *whole* deal.

Ben comes running back, demonstrating authority in all directions, hooting, jumping around from trunk to trunk, shaking branches, running in big circles. His performance sends more onlooking chimps up trees, with much agitated *waah*ing and *aah*ing.

Satisfied with himself, Ben sits down with a regal air.

Unseen — not far — there's a loud altercation. Gerald races in screaming and extends a hand to Ben. Ben clasps it quite as though shaking hands. This greeting says, "We're okay with each other." Ben immediately raises an arm, and they begin grooming.

To be honest, it's all getting on my nerves. The males' demands for acknowledgment of rank, the intimidated screaming and submissive grunting, the young and the females caught in cross-firing ambitions and in the obsessive insecurities of high-status males —. It's wearing on me. It doesn't just waste everyone's time; it's *oppressive.*

"They make their lives much more unpleasant than necessary," I venture.

"Yeah," sighs Cat.

"It's like living in a gang," I say.

"More like the Mafia," Cat suggests.

The mass of chimps lead lives of raucous aspiration. They roil their leafy sea of tranquility into waves of self-generated storms. Any halcyon moment can flip to all-out riot.

Virtually all the problems chimps create for themselves are caused by male aggression driven by male obsession with male status. Caught in a social web of inflicted ambition, suppression, forced respect, coercion, intergroup violence, and episodic deadly violence within their own community, chimps are their own worst enemies.

Birds and mammals that I've studied defend their nests, their mates, their territories. Frequent chases, occasional fights; it's all part of life. I get that. But no other creature that I've studied has impressed me as *vain.* Chimps are vain. And the vanity of chimps — is male.

Quite simply, it's too familiar. You can see in so many ways that chimps are our close relatives. Chimp male status is gained, lost, and enforced through threats and violence. With chimps as with humans, male passions don't just waste everyone's time; they waste the potential for better-quality time.

It's not just that chimps remind me of human beings. They remind me of something worse; they remind me of certain men I've known. The tedious ridiculousness of chimp male dominance highlights the tedious ridiculousness of the dominance obsessions of too many human males. It's the testosterone talking. Chimps' displays of maleness for the sake of maleness make them — and everyone else — victims of male hormones.

Testosterone, oxytocin, and cortisone are three main hormones that help create mood and motivation. They help govern aggression, bonding, and stress. Most other animals share these same hormones. Chimps have them. We have them. That's one reason we easily recognize drives and emotions similar to our own in many species. We all know the feelings. The issue, with chimps as with humans, is a matter of emphasis.

The worst case of misplaced violence Cat has seen involved the older male named Duane. Duane had always been a strategist. When he lost the rank of alpha, his sexual strategy changed. No longer able to monopolize females, he began taking females *away* from the group for extended periods. Such sojourns are called "consortships." The strategy seems to be "If I can't get preferred access to females by dominating all males, I can get access by dominating one female at a time and isolating her."

A consortship might begin just as a female is beginning to come into estrus, a week before she is ovulating. Duane had recently gone away with an older, experienced female. When he'd first invited her to leave with him, she had not responded. After he'd shown mild aggression, she'd gone with him. She might have felt coerced.

Soon after that, Duane started inviting a younger female named Lola. She responded as though the invitation was for sex, but when she put herself in position for copulation, he didn't copulate. Sex wasn't what Duane wanted at that moment. Duane wanted her to follow him away. Signals of sexual invitation are quite similar to those requesting consortship: gestures such as leaf clipping and branch shaking — but without an erection.

She didn't seem to understand. She'd come over to him and present for sex; he wouldn't copulate, and she'd leave to rejoin the group. Frustrated, he would become aggressive. He'd again invite, and she'd again present for sex. He'd again not engage in sex. She'd again leave. He'd become more aggressive. This miscommunication recurred over several days.

Then things got out of hand. Off where other males couldn't hear her screams, Duane attacked Lola.

"He'd beaten her up incredibly badly," Cat recalls with obvious pain. "She tried using me as a shield. We don't interfere in their behavior. And I didn't, but I had tears streaming down my face." Lola climbed a tree. Duane pursued. Then, from forty feet up, Duane threw Lola out of the tree. "It was *completely* terrible that he did that," Cat says. When other males finally did arrive, Duane ran off.

"A large amount of our whiskey got used up that evening," Cat avers.

By morning, Lola had died.

In the logic of genetics and evolution, it makes *no* sense for a male to murder an adult female. Complex minds are capable of miscommunicating, getting frustrated. Complex minds — perhaps *only* complex minds — are capable of becoming irrational. Maybe the explanation is something horrendous: that a female who refuses to be his vessel becomes worthless to him, so there is no downside to venting a murderously frustrated rage.

"Do they ever just horrify you?" I ask Cat.

"It probably helped that I got to know them before" — she hesitates, gives a nervous laugh — "ha . . . before they turned evil. But seriously, by the time we got to that bad stuff, I was already very much a chimp person at heart, so despite it, I'm still here." She's quiet for a moment, then adds, "I guess it's like, you don't get to choose your family members. They are who they are. You love them despite their flaws. Chimps do feel like family. I am

424

going to spend my life with them. And that brings the good *and* the bad. I mean —." Cat pauses, trying to summarize her feelings. "There are horrible times in the forest. But the good times — are good enough."

Of course, male chimpanzees aren't the only large primates who on rare occasions coerce or even kill a female they demand to have a relationship with. Chimps horrify and delight us because we recognize in them parts of ourselves. We see in them aspects of our own passions, and so they hold us in fascination. We cannot look away. So much of what is uncomfortable for us in watching chimps is their excruciating similarity to us. We dread what is sinister in them because we resonate with it; it makes us feel just a little culpable. It is too close to allow us the distance that would let us feel cleanly unindicted.

Chimps seem a cautionary tale. Their proximities unnerve us, and so does their honesty. We tend to deny the dark and unflattering aspects of human nature. Chimps deny nothing. They are all too unfiltered about who and how they are. The chimps are brutal, imperfect, often insensitive. So are many people. Wrote Frederick the Great in 1759, "In every man is a wild beast. Most of them don't know how to hold it back, and the majority give it full rein when they are not restrained by terror of law."

425

Violence *within the community* is a defining quirk of chimp life. Humans share that kink. Chimpanzees and humans are the only primates that make tools *and* hunt in groups for meat, *and* engage in community-against-community warfare, *and* sometimes kill individuals inside their own social groups whom they know well.

Killer whales, sperm whales, elephants, and wolves showcase pinnacles of animal weaponry, power, and intellect. But they do not kill members of their own group. The group exists because all benefit through mutual cooperation in food finding, child support, and defense. When a chimp forms an alliance with another chimp, the basis of the alliance is always that, in order to win in chimpanzee society, someone must lose. The difference between those other animals and chimpanzees is like the difference between musical groups and team sports: in one system, everybody wins; in the other, someone must lose.

Why can't chimpanzees just be nice? Why can't we? Because we're not bonobos. And that's just bad luck. We are as closely related to bonobos as to chimpanzees. Among chimps, humans, and bonobos, some of the thinking is similar, some of the fears and ambitions and emotions the same. But in bonobos, the highest-ranking individual is always a female. Dominance by females

426

makes a big difference for them, because female dominance differs from male dominance. In bonobos, the deep internal impulses for creating and maintaining the peace surpass those in both chimpanzees and humans. In the bonobo brain, the areas that perceive distress in other individuals and those that dampen aggressive impulses are enlarged. "The bonobo may well have the most empathic brain of all hominids, including us," writes Frans de Waal.

Bonobo females form alliances to keep male aggression in check, preempting violence. Among bonobos, fighting is rare; murder is unheard of. By comparison, chimpanzee males form alliances and maintain dominance with violence. Chimpanzee males use violence to entitle themselves to sex; bonobo females use sex to deter violence. Chimpanzees resolve sexual conflict with power. Bonobos resolve power conflicts with sex. Chimpanzees limit sex to one posture and mainly one function; bonobos have never seen genitals they don't like. Bonobos welcome and frolic with strangers, flirt rather than hurt, make love not war. In experiments, bonobos unlock doors to eat with strangers, even when it means being outnumbered by members of another group. No chimp would ever do that. Chimpanzees fear and attack strangers. Bonobos will even let a stranger get into a food-filled room that they themselves cannot get into.

Bonobos have been said to have a threefold path to peace: little violence among males, between sexes, and among communities. The only person who has studied *both* free-living chimps and bonobos, Takeshi Furuichi, has noted that the difference is striking. "With bonobos everything is peaceful," he has said. "When I see bonobos, they seem to be enjoying their lives."

Led by alpha females, bonobos do quite nicely. Why do chimps have alpha *males*? No one knows with certainty how each species got to be the way it is. The full answer probably isn't simple. But perhaps part of the answer is, chimpanzees have alpha males because: they just do. Perhaps male aggression created a self-perpetuating system chimps simply got stuck with. Bonobos don't have that system. Gorillas don't have it. Orangutans don't.

Gorillas live in small groups with one adult male. Different gorilla groups move about the same area, simply avoiding each other. Orangutans generally mind their own business in relative solitude. Compared to gorillas, bonobos, and orangs, we humans are more socially aggressive, more violent, more vain, more political, more irrationally driven, more prone to escalation through sheer emotion.

We aren't "like apes." We are like *chimpanzees.* Chimpanzees are obsessed with dominance and status within their group; we are obsessed with dominance and status within our group. Chimpanzees oppress within their group; we oppress within our group. Chimpanzee males may turn on their friends and beat their mates; human males may turn on their friends and beat their mates. Chimpanzees and humans are the only two ape species stuck dealing with familiar males as dangerous. A gender that frequently creates lethal violence *within our own communities* makes chimpanzees and humans simply bizarre among group-living animals. Chimpanzees don't create a safe space; they create a stressful, tension-bound, politically encumbered social world for themselves to inhabit. Which is what we do. This behavioral package exists *only* in chimpanzees and humans.

We often welcome and aid strangers, but we also fear and harm strangers. We feel most friendly toward those who share our group identity, and we obsess about emphasizing group differences with flags, teams, club insignias, funny hats, special songs, and so on. More important than any insights chimpanzees may elicit into "origins of human tool use" or of "language," chimps may hold clues to the genesis of human irrationality, group hysteria, and political strongmen.

All great apes recognize themselves in mirrors. Can we recognize ourselves reflected in the triviality of chimps' continuous violence? Do they have to be living lives of self-induced stress? Other species show that they don't. We don't. Yet we do.

Of course, many humans achieve mastery over their impulses, and some, magnificently, even work for a better world. That's also who we are. Yet if you look at what we do to the rest of the living world and, too often, to other people, our better angels don't seem generally in command. That's why we have such problems.

Why — for chimps and for us — must this be so hard? Something better is possible. Other creatures have done the experiment for us. All the other apes, plus elephants and wolves, orcas and sperm whales, lemurs, hyenas — they show us a path to being better people. Their conclusion: it isn't necessary to be so nasty to those you know, those in your own community. Being nice, being supportive — that can work.

In elephants and killer whales and some others, as we've discussed, status comes with maturity; no violence is involved, and no elder gets deposed. Individuals ascend to leadership with the wisdom of age, because their knowledge is valuable. Most such species happen to live, as do bonobos, in female-led groups. Their societies emphasize social support. In various species (including humans), females surpass males in offering consolation after something upsetting happens.

Although other species show us that there are different, better, less stressful, less obsessive-compulsive ways to earn leadership, the way it is for chimpanzees is simply the way it is. This must be a cautionary tale for those of us who'd like to move humankind off our square. Our inability to conquer violence is frustrating. But we have the ability to recognize that it's a problem. From that capacity springs our eternal hope. Chimpanzees seem to have trapped themselves in a more violent society than was necessary. The question for us is: What keeps them there; what keeps us there?

It's been another difficult day of hard bushwhacking through thick vegetation, more than once following chimps on our hands and knees. We head for the lowering sun, casting lengthening shadows. On the long walk home to camp, there's plenty of time to talk.

"Why — ?" I am asking Cat. Why are chimpanzees so violent? Why are they so obsessed with status, with an enforced hierarchy, with male advantage? When other species have found a more peaceful path, why do chimpanzees create so much nastiness? I tell Cat that as I've watched the chimpanzees daily for weeks now, I've noticed that both

431

chimpanzees and humans make everyone's lives less pleasant by the self-imposed, self-inflicted burden of violent impulses, impulses that other hominid species generally avoid. Pleased with my insightfulness, I ask, "Don't you agree?"

She doesn't.

"I'm afraid you've gotten a skewed view of male aggression," Cat offers diplomatically.

She says I'm being too harsh — on us humans *and* on chimps. "What you're seeing isn't what the majority of chimp life is like," she informs me. Cat explains that because it's the dry season, the fruit trees are loaded with food, so more chimps are spending more time together in larger groups. Larger groups spark more interactions. More excitability. More females are in estrus. "Plenty of food and plenty of action," Cat sums up. In the rainy season there's less food, and it's more scattered. Chimps spread out and parties are much smaller. Life is quieter then.

She says, too, that we've been following males. Males go where there are other males, and males generate most of the drama. If we'd been following mainly females, we'd see where the peace resides.

So I'd gotten the impression — an impression that the chimpanzees themselves perpetuate — that the males are lording their testosterone-driven advantages of size and aggressiveness. This is in many ways an impression so partial as to be mainly false. Most individual males are never in charge of much of anything. And a male who is "in charge" has so much to worry about from others that he is in large measure oppressed and imprisoned within his own status. Overall, many supposed "advantages" of being born male might be largely illusory.

Chimpanzee top rank is always male, and the top-ranker becomes father to the most offspring during his tenure. Most biologists write that maximizing the number of offspring is "advantageous" for an individual. But that's only one way of looking at it. Yes, there's a breeding advantage to achieving high status if you're born male. But few males reap the benefits — often marginal — that come with the stresses and the sometimes-fatal fighting for status and to maintain dominance.

So another, fairly heretical way of looking at it is that providing the offspring that the gene pool and the species require puts individuals at a personal *dis*advantage — especially males. Obsessed with status, most male chimpanzees suffer from ills of ambition. Male infants are more likely than females to be killed, by males. Males are more likely to get fatally injured, either in a dominance fight with chimps in their own community, whom they've lived with for decades, or in territorial clashes.

Waibira currently has an incredible thirty adult males. That means that any particular Waibira male faces very poor odds of achieving top or near-top dominance, which, in any case, would last a few years at most. And although the rare individuals who become alpha males sire more offspring while they're alpha than any other *individual* male, their edge is no monopoly. In some populations, the alpha sires only about one-third of all babies.

Compared to males generally, females come out ahead. Most males father few or no offspring. Essentially all adult females become mothers. "The most kids a male here has fathered is seven or eight," Cat says. "We've got females in Waibira who've had seven babies, and Kalema, in Sonso, has given birth to eight — more than any chimp known in East Africa." She concludes, "If you're talking about the number of offspring, I don't think there is any advantage to being even a high-ranking male compared to an adult female." For most males, being born male appears to be a disadvantage.

The glue of chimp male allies is often an underlying competitive dynamic. They can be close supporters or shallow schemers. Commonly they are both. We could call them "frenemies."

Males "can be tremendously friendly and mutual with one another, and move around together in seemingly perfect harmony," writes primatologist Vernon Reynolds, a pioneering observer of Budongo chimps in the 1990s. "Small fights are easily contained. . . . But when real trouble breaks out there seem to be no limits to the aggressiveness of chimpanzees."

In a chimp community, it's normal for adult females to greatly outnumber adult males, because males die at a much higher rate, largely from infected fight wounds. That's how it is in Sonso — about two females to every male. But Waibira, Cat says, "is radically different; it's just boys, boys, boys." Waibira has a nearly even ratio of adult males and adult females, around thirty of each. You'd think a lot of males would mean a lot of fights. But it's the opposite.

"Because there are so many big boys, getting into a fight is now extremely risky," Cat explains. "So in Waibira it's more like 'Use your words, not your fists.' " Life for males in Waibira, she notes, is "much more gestural." If a female is in estrus in Waibira, all the males will gather around her, all in competition. One will have sex with her. Then another male will come and shake his hand, or embrace or groom him for a little bit, then walk past him and have sex with her. "If they were Sonso chimps, and another male even dared try to copulate, there'd be a fight. Waibira

435

chimps seem to have a very different style of male interaction."

Cat had used the term "leadership style." We don't often use the word "style" for non-human beings. But that's only because we're not tuned to non-human customs. A style is a way of doing something that ripples through a community. Chimps have styles. Perhaps it could be said that cultures are made of styles. Perhaps Waibira's elevated use of gestures to defuse potential violence is not just a conse-quence of the high proportion of males; perhaps it helped *cause* it.

Regardless, I'm seeing that there are great ironies within chimpanzee life. It *looks* as though being the top male is what everyone should want, but the benefits are largely il-lusory. It *looks* as though males, with their blustery displays, are obsessed with status. But some, calling little attention to them-selves, subtly opt for a peaceful path outside the politics. It *looks* as though males create unnecessary violence. But very ironically here in Waibira, the heightened potential for un-necessary violence posed by the community's extraordinary number of adult males appears to have brought a cultural accommodation — a restraint — that makes violence avoid-able and keeps the peace.

PEACE:
FOUR

On Long Island, New York, I have an office in a house that goes back to 1730. When the house was built, no European had more than an inkling of the existence of any of the creatures now commonly called the great apes: the large, tailless primates we know as chimps, gorillas, orangutans, and bonobos. Back in 470 B.C.E., people from Carthage had reported hairy creatures who threw stones at them in what is now Sierra Leone. Centuries later, around 145 B.C.E., a Carthaginian temple illustrated "satyrs" of the upper Nile, likely chimps. In 1598, the Portuguese noted creatures that were probably chimpanzees in regions of the Congo or what is now Angola. In 1641, a Dutch anatomist, Nicolaes Tulp, described an ape — of some kind.

Not until 1739 did the French artist Louis Gérard Scotin draw the first good likeness of a chimpanzee. In 1740, George Buffon obtained a live individual of an unknown species and described it: "A monkey as tall, as strong as Man, as passionate for females . . . a monkey that can carry weapons, that can use stones to attack, and clubs to defend himself and . . . has a kind of face with traits close to Man." Unveiling his *Systema Naturae* in 1758, Carl von Linné included a creature he named *Satyrus tulpii* — the chimpanzee. (The species' current name, *Pan troglodytes,* offers no particular flattery. Pan was a hairy Greek god, and since Aristotle and Herodotus, a mythical race of cave dwellers were called Troglodytes.) In the 1860s, people like Thomas Henry Huxley and Charles Darwin posited that humans and great apes were close relatives.

In the early 1900s, Wolfgang Köhler gave captive chimps a bunch of out-of-reach bananas, some boxes, and a few sticks. He described their flash of insight when they realized that they could stack the boxes, then knock the bananas down with a stick. In 1933, the bonobo was identified as a distinct species. In the 1930s, free-living chimpanzees got a first brief behavioral study. In 1961, the United States sent a four-year-old chimpanzee from Cameroon — hunters had likely caught him by shooting his mother — into orbit.

By that time, the era of behavioral and medical laboratory experiments was well under way. It wasn't until 1964 that Jane Goodall's emerging, widely covered reports began giving the world any idea about who chimps are and how they live. She and the Japanese researcher Toshisada Nishida, in 1960 and 1963, respectively, had undertaken the first two long-term studies of free-living chimpanzees, both of them in Tanzania, adjacent to Lake Tanganyika. In 1967, a study of blood proteins proved that African great apes are humans' closest living relatives. Many people had been unconvinced by the obvious similarities and needed proof. Many others would not accept the obvious *or* the proof.

Jane Goodall's simple report that East African chimps use sticks to probe for termites shook humanity's sense of human uniqueness. We'd *defined* ourselves as "the toolmaker." But, in fact, news of chimpanzee tool use had been overlooked for more than a century. In 1844, a missionary to Liberia had reported wild chimpanzees cracking nuts "with stones precisely in the manner of human beings." None other than Charles Darwin had written, "It has often been said that no animal uses any tool; but the chimpanzee in a state of nature cracks a native fruit, somewhat like a walnut, with a stone." Even his mention was forgotten.

Goodall's rediscovery that chimps use tools,

published to the attention of millions of *National Geographic* readers and watchers, finally sank in. Goodall's mentor, Louis Leakey, compounded the import of her finding by booming a shuddering, ego-shattering assertion: "Now we must redefine 'tool,' redefine 'man,' or accept chimpanzees as human."

Leakey's radical remark was unusually open-minded and, at the same time, naïve, because we now know of various toolmakers and users across a wide spectrum: monkeys, sea otters, and wrasse fish, not to mention vultures who use stones to crack things like nuts, shellfish, and eggs; herons who use bait to lure fish; and finches and parrots who make insect probes and (in captivity) craft food rakes, or use water to soften food, or fill tubes with water to float food into reach. There are dolphins who use sponges to glove their snouts, for protection against spines and stings. And even ants who use leaves and soft wood to sop up liquid food, and wasps who seal victims in by dropping pebbles into holes and then pounding them into place. And those are just a few examples.

Most chimps forage for termites by probing with a single tool. But in a place called Goualougo they prepare a two-tool set, using one stick to puncture a termite tunnel, then a finer tool for extracting the termites. In a couple of places, chimps use slender sticks to dip into a swarm of army ants. (They swipe the angry ants off the wand and quickly shove them into their mouths, crunching the insects before they get much chance to bite back.) Others use sticks for digging up tubers. In one group at Tanzania's famed Gombe National Park, young chimps even tickle themselves with sticks.

Chimps also use stick tools to obtain honey. A love of honey and using sticks to get it are two more things we share. When I was in my twenties in Kenya, my new Maasai friend Moses Ole Kipelian showed me how his people dig honey from underground hives of stingless bees. He used two tools: a stout digging stick and a more slender stick. Rather similarly, chimps not far from Budongo, in Bulindi, get honey from stingless bees that nest underground by using digging sticks to access the underground nests, then switching to delicate probes. To get into beehives in hollow branches of high trees, some honey-seeking chimps wield a variety of tools in sequence; these are referred to as a "pounder," "perforator," "enlarger," and "collector." They first pound their way into the entrance with a stout branch, then use thinner sticks to open up access to deep combs filled with honey and larvae, and they probe out the honey with even smaller sticks. At one tree, researchers found about a hundred such tools.

Like humans, chimps make different tools in different places. Chimps here in Budongo Forest — *unlike* chimps everywhere else — don't use sticks or wood tools for getting food. (One researcher noted an exception: "The colobus monkey they've been chasing falls to the ground. Chimps on the ground catch it, pinning it down to the ground with a sapling.") This might be because Budongo is rich in fruit. In 2019, scientists reported a diverse culture of tools and customs in the chimpanzees of the northern Democratic Republic of the Congo. These chimps use long probes of well over a yard to get driver ants, short probes for ants of other types and for stingless bees, thin wands for honey in tree nests, and digging sticks for underground bee nests. They pound open African giant snails, tortoises, and certain termite mounds that chimps in most other regions ignore. And they often make their sleeping nests on the ground. In Congo they make nearly thirty different tools. Chimpanzees' total tool kit as a species includes various kinds of probes, hammers, anvils, clubs, sponges, leaf seats, fly whisks, and other gadgets.

The point is, chimpanzees in different places have differing material and behavioral cultures. It's not just because the four different races live in different regions. Even in adjacent communities where the same plants grow, chimps of one community customarily make longer and wider tools than their neighbors.

After Goodall documented the chimpanzees' stick tools, West African chimps' use of stones for nut cracking was rediscovered. "At the base of several species of fruit trees that drop hard-shelled nuts," wrote one researcher, "the chimps gathered to create workshops."

Only in West Africa west of the Sassandra-N'Zo River, apparently, do chimpanzees crack nuts with stones. For those groups that do crack nuts, which nuts they choose, and how they learn to crack them, and which tools they learn to use varies culturally from population to population. Thus some are said to have "nut-cracking traditions."

Cracking different nuts requires a set of skills because different nuts vary in hardness and shape. Nut-cracking chimps consider the type of nut, then, based on experience, select a tool for the task. For softer nuts, they may choose wood clubs; the hardest nuts require stones. A chimp carefully places a nut into, say, a depression in an exposed tree root, then strikes it with a stone or club. Some arrange a stone anvil. Sometimes a third stone angles the anvil. (Anvils weigh five to six pounds; hammer stones are around two pounds.) A chimp must strike precisely or the nut may jump away when hit. The force used must be sufficient to crack the shell but not pulverize

443

the nut. In some places stones are sparse, so the chimps transport them from one cracking site to another. If the nuts they're headed for are far away, they may opt to carry a lighter wooden implement.

Between the ages of three and five years, an infant watches their mother's cracking techniques. In the Republic of Côte d'Ivoire's Taï Forest, chimpanzee mothers sometimes actually guide their child's nut-cracking efforts until they succeed. By about ten years old, highly skilled individuals can split the whole nut open with just a couple of hammer blows. But some chimps never get the hang of it; they become what researchers describe as "habitual scroungers of others' abandoned half-cracked nuts." Female chimpanzees learn tool use faster and are simply better at it, on average, than males, because male youngsters get more preoccupied with socializing.

Researchers have identified about forty tool-use behaviors culturally learned from other chimps (and in orangutans, nineteen). Orangutans are adept with their own tools, using leaves as hand protectors, face wipes, and seat cushions; making protective awnings over their nests; and even using wood implements to masturbate. In captivity, orangutans have made wire tools for opening door jambs — then hidden them, keeping the befuddled zookeepers mystified about how they kept getting out. I've seen captive orangs string up their own hammocks. I watched one put on a T-shirt prior to napping. After she woke, I watched her tear a strip off a piece of cloth, use it to string wooden beads, knot the ends, and don her new crown. The keeper swore that no one had taught her that. This wasn't just thinking ahead and envisioning an outcome; it was also body ornamentation, a self-concept with an aesthetic flourish. Gorillas rarely use an implement. Nor do free-living bonobos — which is odd, because in captivity bonobos use tools just fine.

Animals are born genetically enabled to perform or learn certain behaviors, including the traditions of our tribe. But not everyone, everywhere, learns everything. Although chimpanzees in East Africa do not crack nuts, East African chimps brought into a sanctuary with skilled West African nut crackers easily learn by observation how to use stones as hammers and anvils. The East Africans have the *capacity.* They don't have the *custom.* They can acquire the custom. That's what culture is about.

Culture enables adaptation at speeds far faster than genes alone can navigate hairpin turns in time. Necessity is the maternity ward of invention. Because the Sonso community borders cropland, crop raiding has become part of the Sonso chimpanzees' culture. Crop-raiding chimps go for mangoes, maize, sugarcane, papaya, anything they deem tasty.

If people are in view, the chimps will not venture out into the open. When the farmers take their produce to market, the chimps raid. In some places, chimpanzees override their instinctive fear of night and go crop raiding *nocturnally*. Chimps casting moon shadows while foraging is a new thing under the sun, another cultural innovation.

Flexibility that becomes shared habit is called "custom." Custom learned through generations becomes tradition. Traditions make up culture. You know there is culture if not everyone is part of the same culture. A culture can be a package of traditions, a repertoire of behaviors, skills, and tools characterizing a group in a place. As researchers have written, "Distinctive tool-using traditions at particular sites are defining features of unique chimpanzee cultures." Cat's colleague Andrew Whiten wrote, "Chimpanzee communities resemble human cultures in possessing suites of local traditions that uniquely identify them." He says they have a "complex social inheritance system that complements the genetic picture."

Today after watching chimpanzees rouse at dawn, move up into food trees, and spend some hours eating and resting, we've followed them on their daily trek to the water holes. They get right down to business, and the drinking is quiet. Some chimps sip straight from the muddy puddles, both hands down and the body in a deep bow. Others grab an adjacent sapling and use it to spare their hands from getting dirty as they bend for a drink. The young watch their elders, absorbing how to be a chimpanzee.

It's called the dry season for a reason, and thirsty chimps crowd the shrinking drinking spots. However, one adult, Onyofi, has innovated a way to get clean water. Using her left hand and some determination, she rakes a shallow well and waits while it fills. (Onyofi's name means "finger." Her right index finger sticks straight out from a paralyzed hand.)

Everyone around has been watching Onyofi, interested in what she's doing. One of the babies imitates her digging. He doesn't seem to know why, other than that he's doing something a grown-up is doing.

Next, Onyofi gathers a handful of leaves, chews them into a wad, takes the wad from her mouth, and dips this "leaf sponge" into her well. She brings it dripping into her mouth, presses the water out with her tongue, savors the private drink her work has earned her, and repeats.

Twenty-year-old Tibu also digs a shallow well and makes a leaf sponge. Tibu's little one begs for her mother's leaf sponge, poking it and touching her own mouth.

Young chimps often sponge-drink alongside

their mothers. Hardly surprising; they do *everything* alongside their mothers. As with much learning in traditional human societies, most chimpanzees learn by simply watching.

Even in many human cultures, children learn by watching rather than being taught. For instance, in his memoir, Native Alaskan Athabascan elder Henry Beatus Sr. recalled, "I see my grandmother cut fish so I use my little knife. Pretend that I am cutting and drying. . . . I just copy her. . . . She didn't help me. I just watch and get the idea to follow."

What must a chimp learn by watching other chimps? Put it this way: they must learn *everything* — starting with who they are, defined by with whom they belong. Almost never has a captive chimpanzee successfully returned to nature. Chimpanzees raised by humans are as unprepared to cope with — or be accepted into — free-living social situations as we would be if we were released into an indigenous territory in the Amazon rainforest. Released apes usually starve or get killed. Their long childhood, like ours, is for learning how to become who they will be. They must *learn* how to be normal. Theirs is a wild life that isn't what we thought it was; it's a cultured existence.

Onyofi, Tibu, and some of the other chimps make basic wad-style leaf sponges for drinking. Sixty years of studies tell us that all chimp populations make leaf sponges. Alf

448

uses a common variation: he folds big leaves into accordion-style sponges. *Moss* sponges — something new — seemed to suddenly appear in 2011. Used by just a few chimps, moss sponges are faster to make and hold more water. They're a cultural improvement. Yet the use of moss sponges is spreading slowly, with most leaf spongers sticking with the tried and true.

Watching Onyofi we see — and the other chimps also see — that it's possible to dig a shallow hole in the wet mud, wait for it to fill with water, and have a private drink — yet only Onyofi and a couple of others regularly dig holes for water.

"They've seen someone dig for water and they've drunk from her well, but —. They're *so* conservative." Cat sounds almost exasperated.

Given this conservatism, it's surprising (and ironic) that where chimpanzee territory abuts farms, the chimps have rapidly learned to exploit crop varieties with which they've had no prior experience, such as guava. Cat asks, "Can they generalize with rules like 'In the forest avoid strange food' and 'Outside the forest, people foods are safe to eat'?"

I suggest, "Maybe it's like growing up watching people eating Marmite."

Cat laughs because the analogy is a pretty good one. In Scotland, where Cat lives when she's not in Uganda, some ordinary, even

449

celebrated foods — Marmite and haggis notoriously among them — can seem, to outsiders like me, strange and inedible. Chimpanzees in the Waibira community sometimes catch and eat both red and blue duikers. But if Sonso chimps catch a blue duiker, they abandon it. Their cultured, wasteful pickiness *does* seem crazy.

In many human cultures, people eat eggs, fish, insects, or rodents. In many other cultures, some of these foods are not considered edible. Years ago, when I lived for a few weeks with some Maasai people in the Loita Hills of Kenya, I watched them drink fresh blood streaming from the necks of cows, but they believed they'd get sick if they ate eggs or fish or the heads of the goats they slaughtered — all of which, in Europe and many other places, is food. Most Westerners won't eat grasshoppers or rats — both of which I have seen displayed in food markets on other continents.

Chimps who won't eat guava, say, simply because they've never eaten it might seem crazy to those of us who know what guavas are. But people in the tropics eat many fruits that visitors pass up simply because we've never seen them before, don't know how they taste, and have no idea what to do with them. Durian is considered "the king of fruits" by some, but to those who have not learned to appreciate it (I have not), its mere smell can seem revolting. Some fruits are poisonous to humans, so hesitance about the unfamiliar is wise. Food has risks. What is and isn't food, and how we eat, is learned. Food is cultural. So before faulting chimps for being too conservative about food, look in your refrigerator. Anything *un*familiar in there?

Again, little chimps begin learning their culture by watching their mothers. As far as "What's food?," they learn "This is good" and "This, avoid." They learn where food trees are, when they ripen. In a food tree, an adult might be surrounded by a ring of juveniles closely observing them as they eat, often right in their face.

And as we've seen, all their social etiquette — who gets respect, who gets snubbed — must also be *learned.* A male named Zig had lost a hand to a snare and an eye to a fight. He was smaller and weaker than his age-mates, and he grew into young adulthood on the periphery of the groups, often ignored by the others. One day Kalema, with her young infant Kirabo clinging, walked right past Zig to join several others who were grooming. But Kirabo slid off her mom's back, ran back to Zig, and gave him a little kiss and greeted him. For a little chimp who was just learning, Kirabo seemed to think, "We greet everybody." "It was such a sweet moment," Cat recalls, "to see Zig get respect." But

451

Kalema quickly came and whisked Kirabo away, as if to say, "We don't socialize with *him.*" "It showed me, you know, they're not *born* political. That side emerges as they learn to navigate the social world. Their more basic instinct is to be curious, make friends with everyone, and be positive," says Cat. "You learn the rest later."

Among animals generally, the importance of mothers in learning is underappreciated because, well, who has time to watch thousands of species growing up? A few people *have* watched a few species, and often: mothers are crucial.

Grizzly bear expert Barrie Gilbert got to know a female who fished for salmon at Alaska's McNeil River by standing on a couple of particular rocks with her paw cocked in a certain way so she could swat leaping fish. Her youngster learned exactly the same stance. Black bear expert Ben Kilham has raised and rewilded hundreds of orphaned cubs. When I visited him in New Hampshire one cub-filled spring, he explained that his starting premise is that the cubs' genes give them the body, senses, intelligence, and psychology needed to survive in their world — but they require an opportunity to learn *how* to use that endowment. That opportunity is their mother. The mother takes them throughout a complex physical and social environment, keeping them safe while introducing them to all the foods, situations, and dangers they'll have to access, confront, and respond to.

"If you go out walking with cubs," Ben explained, "you see that they want information." Some plants are poisonous; how do they learn what's good? Ben did an experiment. He knew the tiny cubs he was out with had never eaten red clover. "So I found some, bent down, put it in my mouth. They rushed over and stuck their noses in my mouth, smelling. Then they immediately went searching for whatever smelled like what I was eating, and found some red clover and ate it. This is how they socially learn foods, from their mother." Foods vary from region to region, so such food traditions are an aspect of bears' culture. "We had a young cub, Teddy," Ben recalled, "who hadn't eaten jack-in-the-pulpit, even though it's a food that bears eat regularly." An older bear named Curls came around and was eating jack-in-the-pulpit. "Teddy followed her, sniffing in all the holes where Curls had pulled a plant. Then he found some growing and started eating it. This is all *social* learning," Kilham noted, "though books tell you bears are solitary."

Perhaps the most bizarre example of young picking up adult culture from parents — and, therefore, the most instructively eye-opening — is the very strange case of the mallard duckling who was adopted by loons and did some very un-mallard-like, purely loony things. Mallards *never* ride their parents' backs (loons do, and this adopted mallard did); mallards *never* dive underwater (loons do, and this adopted mallard did); mallards *never* catch fish (loons do, and this adopted mallard ate the fish its loon parents fed it). When a nice, normal loon family has a nice, normal chick or two riding around, diving, and eating fish, we assume chicks ride their parents by "instinct," dive by "instinct," and eat fish simply because that's loon for "dinner." It takes a wayward duckling in an alien family to give us a whiff of how much cultural learning goes on, and how much flexibility exists each step of the way, in becoming wild.

Bears to loons and beyond, youngsters of many kinds watch mothers and elders, learning what's good and how we do things. Just a few years ago, many behavioral scientists considered learning by watching to be "exclusively human." But even informal experiences with other species — for instance, seeing young dogs model their behavior on older dogs — reveal widespread tendencies to copy.

We once hand-raised an orphan raccoon, who lived under the back porch. She would climb the screen door and manipulate the doorknob when she wanted to come in, though she had never opened a door that way. We'd never made any attempt to demonstrate this. (We did not want her to let herself in!) She was imitating what she'd watched us do. Goslings who have observed a human opening a box focus their attention where the human hand contacted the box. After a dolphin who'd learned tail-walking during temporary rehab was released, wild dolphins started copying the trick, for fun. (Captive dolphins copy humans more than apes do; that's impressive, considering that dolphins have no arms or legs.) In what was called "a thorough demonstration of cultural diffusion," bumblebees were trained (yes, some insects can learn) to pull a string to drag an artificial flower from under a cover and drink from it. Bees who merely watched a trained bee mastered the technique. Further, over the three bee generations that the experiment ran, two-thirds of the new bees — literal newbies — learned the technique from others who'd originally learned just by watching. Culture in an insect.

So, there's plenty of copying. *Teaching* is different. Teaching occurs when a skilled individual takes time away from what they are doing to help a less-skilled individual learn or improve the skill. Teaching is rare in other-than-human species. So rare that, as late as the 1990s, scientists were still asking, "Is there teaching in non-human animals?"

Chimpanzees don't actively teach; that's the official line. Except — sometimes they do. After one six-year-old chimp picked up his mother's hammer stone and nuts and placed a nut on the anvil, his mother picked up the nut, cleaned off the anvil, and placed the nut back in a better position. The young male pounded open the nut. In the Republic of the Congo, mother chimps gave their children their tools or broke their own stick probes and handed them one-half. This cost the mother time and efficiency in her own tool use and eating; that's the price and the definition of true teaching.

Mother chimpanzees as well as gorillas, rhesus macaques, baboons, and spider monkeys sometimes encourage small babies to walk and to follow. One scientist described watching yellow baboons: "A mother began to take a few steps away from her infant, paused, and looked back at the infant. As soon as the infant began to move toward her, she again moved slowly away." This was repeated every few steps, until the baby got good at following along.

When spotted dolphin mothers hunt alone, chasing a fish usually takes less than three seconds. But when foraging with youngsters under three years old, they often let the chasing go on for half a minute, sometimes repeatedly releasing and catching prey to prompt the young to participate. Various cats, from house cats to cheetahs, jaguars to tigers, release live prey near their young ones. Orca (killer) whales sometimes stun prey with their tails for their young ones or toss prey to them. In a couple of places, killer whales help juveniles practice beaching, a specialized skill they'll later use for catching sea lions on shorelines. It can take years to master this beaching skill, from the time the young first start playing at it, before age three, to their first successful solo beach grab, at age six. One youngster had a "supermom." During that young one's first successful sea lion capture using the beaching technique, the youngster received an assist onto the beach from Mom and then, after the snatch, assistance back into the waves. That's some pretty fancy, truly cultural teaching.

Like human mothers, chimpanzees take food away from their infants if what they're trying to eat isn't part of their community's diet, and monkeys discourage their babies from ingesting things known to be poisonous.

Far from being "exclusively human," social learning, even teaching, enters the lives of a wide array of animals, all closely watching, keenly primed to copy what works.

So if you're an infant chimpanzee, you go with the group on a particular trail to a particular tree. Now you know where that tree is. You'll learn what the food looks and smells like when it's ready. You'll feel the season when it's ripe. And you'll know where water is, and how to make sponges that are especially helpful in the dry season. Alone, you would not know. But now you know, because you got led here as a child. You saw others do it. You've learned "how we do things" — your culture.

To learn is to become. Some animals cannot "become" without a social group. A honeybee cannot be a honeybee without being a member of a colony in a hive. A human in isolation cannot be human. A chimp alone is not a chimp; chimps need chimpanzee society. Social animals must live in, be part of, and help create their appropriate social context or they cannot be, or learn to become, who they are.

Young chimpanzees learn easily. But like humans, chimps tend to become set in the ways they've learned. Then they want only conformity.

The safe bet is: do what is working for everyone. If your mother learned what a hundred generations learned, and she taught you what foods are good and what plants or fruits or places to avoid, experimentation could cost you dearly. I once watched a long line of wildebeests traveling single file across short grass to a water hole. Their route was so circuitous it seemed stupid. The whole line of them took a long detour, went around a small tree, and continued to the water. Everyone slavishly followed the footsteps of the one in front of them. Why single file, why the long detour — why not cut straight to the water? Answer: the individual immediately in front of you did not get attacked by a lion; why risk stepping out of line? Years later, I spent several nights watching a water hole in Namibia through an infrared scope. All the antelope and zebras that thronged it by day evacuated at night. But one night, out of the darkness, a lone springbok very cautiously approached the water. When it drew alongside a fallen tree, a lion leapt over the log and another nonconformist bit the dust. Genes for conforming stay in the gene pool.

Even when a chimpanzee does innovate something novel, that individual will often revert to the norm for the group. When an immigrant female chimpanzee brings skills from her birth community into a new group, the residents seldom adopt them. Instead, the immigrant tends to abandon the skills she brought and conform to the behaviors typical of the new group. ("In every work of genius," Emerson observed, "we recognize our own rejected thoughts.") When female western chimpanzees leave their birth group to join another community, they sometimes adopt a less efficient nut-cracking technique, conforming to their new social group. Rather than promote cultural advance through a "melting pot" of all incoming skills, com-

459

munity cultures tend to remain as they were.

"In one sense, this is the opposite of intelligent," write two leading ape-culture experts, Andrew Whiten and Carel van Schaik. "It could even be described as 'mindlessly following the herd.' " They note that conformity is also "a marked characteristic of human cultural behavior." We humans have "a particularly strong motivation to copy others rather than use one's own knowledge."

But should we be surprised? Human immigrants usually adopt the local traditions. Even in the "nation of immigrants" — the United States — for instance, new residents are often eager to learn how to make the traditional American Thanksgiving dinner. To switch continents, though not species, we could call this the "When in Rome" effect. Humans — especially those who have gained or who stand to gain power from group conformity — pressure others to conform to their religion, language, hairstyles, dress, ceremonial shows of national allegiance, and so on. History and current events are full of oppressive examples. Doing your own thing, asserting your right to self-determination, free speech, liberty, and happiness — it can get you ostracized, even killed.

A clever experiment with free-living monkeys reveals a deep tendency to conform. Researcher Erica van de Waal and her colleagues gave two groups of wild vervet monkeys corn kernels. She had dyed half the corn. For one group, she'd applied a distasteful additive to the dyed corn. For the other group, the distasteful additive had gone into the undyed corn, and the dyed corn tasted good. In each group, the monkeys quickly learned to avoid the corn whose color (either dyed or natural) indicated that it tasted bad.

After the monkeys learned to avoid one color, the researchers stopped using the bad-tasting additive; now all of the corn was sweet. Two dozen new babies were born during the part of the study when none of the corn was distasteful, but all of the babies ate only the color their mothers were eating. The researchers *then* observed individual vervets moving out of one group and into the other. Though they had learned to avoid one color in one group, as immigrants they quickly adopted the preference of their new companions — *eating the color they'd earlier learned to avoid.* This is conformity to local culture. The researchers concluded, "Powerful effects of social learning represent a more potent force than hitherto recognized."

A natural event provided insight when a tuberculosis outbreak killed half the males in a well-studied group of savanna baboons. The most aggressive males died, leaving as survivors an unusually nonaggressive group. A decade later, after all the males who had lived through the outbreak had passed away, the

461

peaceful era persisted. Males living in this group were unusually placid. In this species, adolescent males leave the group they were born into and take up life in a new group. Even though new immigrant males had grown up among typically aggressive male role models in their birth group, when they arrived in the placid troop, they adopted its uniquely pacific culture, including high grooming rates between females and males and a "relaxed" dominance hierarchy.

Social learning would seem to allow individuals to greatly expand on what they would figure out alone. But it also creates *narrowing.* Vocal sounds that everyone is capable of making are called phonemes. Cat points out how the narrowing happens: "A human baby born to Scot or Thai parents has all those human phonemes available. They then *limit* their repertoire to the sounds of their particular language." Social learning can involve contracting the repertoire to just *some* of all possible behaviors. Cat believes that "all social learning is a matter of taking everything you might be capable of and tuning it *down* to a particular way your group does things." It's like that with human cultures and human ways of making a living: a baby has almost universal potentials, but we learn, and we live by applying, the tiniest fraction of all human knowledge and skills.

Part of the pressure to conform is that what

works, works. The observations and experiments above show that doing your own thing might get you eaten by that lion, poisoned by the wrong food, or rejected by potential mates.

In chimpanzees and humans, "conformity overrides the discovery of valid alternative means," write Whiten and van Schaik. Surprisingly, human children are more slavish conformers than chimpanzees. Human children typically copy behaviors exactly. Chimps often grasp the goal and create shortcuts to achieve it. In experiments, when human children watch someone struggling to open something, they usually focus on the parts that the demonstrator has struggled with. Chimps often avoid the problem part and concentrate on a different part. Human children often exactly mimic even useless parts of behavioral sequences, such as knocking on a jar before unscrewing its lid. Chimpanzees often understand and leave out the unnecessary moves. Thus, human children have been described as showing "an extreme form of reliance on cultural convention . . . apparently less rational, emphasizing the extremes of conformity to which our own, super-cultural species is often subject." And we often inflict these extremes of conformity on ourselves and on others.

There are world-shaking implications here, with all that the word "conformity" implies.

Conformity might work fine when the world you're in is stable. Or your culture is fair and justice reigns. But the world we're in is fast-paced and ever changing. What's needed now, among chimpanzees and among us, are a few more nonconformists to innovate adaptations to the changes we are causing.

We've said that culture is "how we do things." But that definition leaves out *innovators,* the most important, the rarest — and the most resisted — creators of culture. In 1953 (before formal studies of free-living creatures' behaviors), one female Japanese macaque, named Imo, started washing sand and dirt off potatoes people had given to her group. Her innovation spread to her kin and to playmates. She became famous as the first-known other-than-human culturally innovative being.

There is no culture without innovation. Intelligence can be understood as the ability to innovate. Yet *culture* is mainly about conformity, consistency, and tradition. Fact is: culture requires *both* innovators, who create some new thing never before learned (and are often ignored and resisted), and adopters, who, by learning, narrow themselves and conform. Interesting. Culture brims with irony. Being conservative is safer than thinking freely. Safer than experimenting and innovating. Yet conformity is, as Whiten and van Schaik put it, "the opposite of intelligent." Without freethinkers and innovators, nothing ever improves, no one adjusts to change, and no culture *ever* arises. We can see this dynamic tension between intelligence and conformity playing out every day.

Wherever chimpanzees live, chimpanzees hunt. But wherever chimpanzees hunt, chimpanzees hunt differently. Even our adjacent Waibira and Sonso communities hunt differently. Monkeys and duikers are equally common in both places. Sonso chimpanzees catch monkeys in more than 90 percent of their successful hunts. Less than 7 percent of the time, they catch a duiker. By stark contrast, Waibira chimps have caught a duiker on fully 60 percent of the occasions researchers have seen them with meat. These two chimp communities live in the same forest and have a shared boundary. Each community includes female immigrants who were born in the adjacent community. What's different between these neighbors is: their hunting cultures. Among groups, there is cultural complexity; within each group, there is rigid conformity.

Chimpanzees at Fongoli, in Senegal, hunt tiny primates called bush babies (also called galagos) by probing tree holes with sticks they've sharpened into crude thrusting spears. Although most hunting in most chimpanzee communities is done by males, in Fongoli the spear hunting is done mainly by females and immature individuals. Hunting with spears is a way females and adolescents can get their own meat, because adult males are stingy about sharing.

Researcher Jill Pruetz recalled for me the first time she observed spear hunting: "I saw an adolescent female carrying a large stick tool, and I immediately knew something was up. I followed." In their scientific report, Pruetz and Paco Bertolani wrote, "Chimpanzees use one hand in a 'power grip' to jab the tool downward multiple times into the cavity." Often the chimps examine or sniff the end of their spear for signs that they've struck their intended prey. One chimpanzee, after detecting a strike deep in a hollow branch, jumped on the branch until it broke off. She then reached into the cavity and pulled out the bush baby; her spear had killed it. Pruetz says that the chimp she followed that first time, named Tumbo, is one of the best tool-using hunters at Fongoli. Not surprisingly, her son Cy is one of the youngest chimps to have captured a bush baby. Pruetz added, "I still get excited every time I see a chimpanzee hunting with tools. There is a lot more for us to discover about how they learn and refine the technique."

Multiple-step toolmaking is called "crafting." Only humans, chimps, orangutans, certain crows, some parrots, and very few others are known to craft tools. Until 2005, the Fongoli chimps' crafting of hunting weapons was completely unknown to humans. While breaking, trimming, and using their front teeth to sharpen a branch into a long dagger (or blunting a slender stem into a brush for getting termites), a chimp must have intent and a hoped-for outcome *in mind.* The researchers wrote that the Fongoli chimpanzees show "foresight and intellectual complexity." They compared them to "early human relatives."

One thing, though. These comments imply that learning to kill with weapons is an *advance.* In one sense, sure, a technological advance. But there are other ways to express superior intellect. For instance, an advance of emotional intelligence: superior empathy. When a bush baby got *into* a pen housing rescued gorillas in Cameroon, the gorillas held and petted it, observed it with total fascination, then very carefully carried it to the edge of the pen and released it. (You can see a video of it on the Web; search for: gorilla bush baby.) It's one of the most impressive things I have ever seen. The capacity to exert care and effort so as to cause no harm — that's *quite* an advance.

As the world's main weapon crafters, humans are well qualified to praise weaponizing chimps for "foresight and intellectual complexity" while condescendingly comparing

467

them to "early human relatives." But I find the gorillas' spontaneous, unschooled gentleness humbling. It shows foresight and intellectual complexity of an elevated kind. It's something we could look up to and hope to emulate. Many humans inflict worse pain and damage than do the spear-thrusting Fongoli chimps. Not many humans are any gentler than the gorillas.

What would the world be like if the "early human relatives" who would evolve into the planet's most dominating, devastating toolmaker had brought with them instead a disposition that walked a more peaceful path, less like chimpanzee, more like gorilla?

PEACE:
FIVE

On this new day, we've been watching Talisker and Ben grooming when we hear a clear chimpanzee call. The chimps don't react. I look to Kizza, my raised eyebrows asking a clear if silent question. He immediately says, "That is Alf." If Kizza can tell instantly, so can Talisker, and so can Ben. No reaction is needed. They also impassively absorb the shrieks of two young ones playing nearby with their mothers, who are of only passing interest in the political lives of males.

Kizza and Cat cannot tell me what about Alf's voice they recognize. But it would be nearly impossible for me to describe in words what I recognize in the voices of, say, my mother or my wife or my closest friends. Our mental voice-recognition system appears to do its analysis subconsciously and just hand to our conscious mind the identification of who's talking. "It's almost easier to describe strangers," says Cat. She thinks and comes up with this: "Well, okay. So, Alf hoots a distinctive 'Ooh.' Lotty's pant-hoot is odd, halfway to a whimper. Nambi, over in Sonso, is a little waily at the end. Things like that."

Another round of calls ends the grooming. We hear a food grunt or two, meaning that the callers are now climbing into a feeding tree. Ben halfheartedly makes several abbreviated hoots, uttering just the ending phrase, more a comment to himself — "Noted" — than anything intended to reach those who are eating.

But then he bangs hard on a buttress root, sending resonant drumlike booms into the forest. "Hear ye, hear ye; I am — Ben! I am coming!"

This helps to set off a round of hooting and screaming from unseen chimpanzees. "Hooting and screaming" are words of convenience. The sounds actually contain a wide range of vocalizations. Labeling a call a "pant-hoot" makes it seem as if all the chimps who make this call are saying the same thing. They're not. One common "hoot," for instance, actually sounds like *Ooh-wó*. True, there *is* a basic pant-hoot pattern. It starts as slow hooting and can have four stages: introduction, buildup, climax, and letdown, or resolution — all the components of a good novel. But a party of chimps eating high in a tree might join in only for the climax, filling the forest with screams, a sudden racket peaking in a few moments sounding to human ears like sheer hysteria.

Some researchers try to sort calls into categories: "scream" or "grunt" or "bark" and so on. There's a "travel hoo" that goes up at the end like a question. There are "dominance screams" and "victim screams." But within those "categories," some utterances are long, short, assertive, or frantic; there's quite a range. Crucially, varying the intensity, loudness, pitch, and repetition can change the meaning and urgency of what's being communicated. The differing intensities of those screams reflect — and so convey — the level of arousal or excitement the caller is feeling. As with humans, chimpanzees sometimes scream when they're excited, either positively or negatively. Screams can signal finding good food. Or they can mean someone is being attacked. Screams can grade into waah-barks, signaling a shift from defensiveness to retaliation, a turning of aggression on the aggressor. What *you* actually hear is a great array of hoots, whoops, wahs, aahs, and screams.

"I used to just think I was really bad at categorizing," Cat admits. "In reality, they all blur."

Some researchers exclude from their analyses all "partial," "incomplete," or blended calls.

"But incomplete and blended calls are a really huge part of their lives," Cat says. "If you discount partial or halfhearted sounds

471

and gestures, you're probably missing a lot of casual communication between familiar individuals. They probably use as much or as little of the sequence as they need."

To negotiate a complex social world with only a few kinds of signals, using just parts of them or changing their intensity can put additional meanings into those signals, thereby enriching communication. The chimps understand what they're hearing. They can glean who's doing what where and to whom. A lot of information comes across the open spans of forest air.

Everybody knows that Alf is eating, because he changed his vocalizations from traveling calls to sounds that mean he is ascending, and when he started climbing he added food grunts into his hoots. We also simply hear that he's now calling from high in a tree, and we decide to head that way. That other big group we hear — they're at the water holes. All of the chimps keep tabs on who is calling and what they're doing. Perhaps knowing who's where, and what they're up to, is all they need to know; maybe it's all they say.

Almost every call draws a chorus of replying screams. The chimps work themselves from the most minor comment to frantic hysteria, seeming constitutionally incapable of getting less than very excited. Their voices sound incredibly emotional. But it's possible that humans overinterpret the emotion in those screams. As quickly as they scream as though in abject terror and mortal danger, they calm down, reconcile, and begin grooming one another and smoothing things out.

"My Lebanese family," Cat offers, "when they're arguing, it's so loud —. Five minutes later, everyone's laughing and we sit down to dinner." I come from an Italian family; I get it. Still, the chimps often *seem* very upset.

High in the tree Alf has ascended, the dark shapes move slowly, reaching out to take what the world is offering them. This tree's seed pods are like large peapods. And, in fact, this towering giant tree is a legume, *Cynometra alexandri,* locally called *muhimbi.* Chimpanzees know it very well. For about an hour we sit beneath the massive *muhimbi*'s vast umbrella canopy while foraging chimps drop empty pod husks into the dry leaves around us.

The bigger males posture as males often do, even while climbing — "I'm here; where's the best spot?" — making sure everyone knows and acknowledges their presence.

Cat had mentioned that the extraordinary number of males in Waibira so raised the risks of violence that it changed the style of interactions here into more of a "use your gestures rather than your fists" kind of community. We'd seen the gestures used in Ben and Talisker's face-saving "scratch-off." And

Alf had used a scratch and an extended arm to invite Gerald's grooming.

As Cat has been discovering, a particular gesture is not the same as a particular human word. It's more like this: clipping a leaf with your teeth *or* shaking a branch — each means "Get sexual with me." Some of the things chimps communicate with gestures include: "Let's make contact," "Give me that," "Follow me," "Let's travel," "Move closer to me," "Move away from me," "Look here," "Stop that," "Climb on me," "Let me climb on you," "Let's groom," "Reposition your body," "Focus your grooming on this spot," "Pick me up," "Let's play" — and there's more.

Chimpanzees and humans appear to have some instinctive, universal calls and facial gestures, such as smiling, laughing, fear-screaming, and, seemingly, grunts of contentment. Cat observes, "When we are eating, we make human food grunts. A good bowl of beans and rice? 'Mmm-mmm.' " It's also what chimps do when enjoying food.

Other gestures, however, are learned socially. Free-living chimps essentially *never* clap their hands or finger-point, but captive chimps learn from humans to clap and point. Captive apes can learn systems involving hundreds of signs in languages such as American Sign or Yerkish, which uses symbols known as "lexigrams."

Why don't they just talk? Chimpanzees and

humans have different versions of the gene called FOXP2, which affects the ability to articulate speech. Chimps lack fine control of their vocal cords. Having something to say and not being able to say it might be why gesturing is a major component of chimp communication.

All apes use gestures. Gesture becomes communication when it's directed at a specific individual, who changes their behavior in response. Sometimes you see the gesturer persisting by repeating or trying something new to get a reaction. Budongo chimps collectively use at least sixty-six different intentional gestures. Of these, about thirty get deployed regularly. Just as each of us doesn't take advantage of all the words in our language, any particular individual chimpanzee uses only around fifteen to twenty gestures out of the regional repertoire. Gorillas have a total repertoire of 102 gesture types. (Apes are not alone in using gestures. Ravens gesture to direct the attention of other ravens. Dogs use a total of at least nineteen gestures to get a point across to humans.)

Chimps gesture frequently. Many gestures are subtle, easy for a beginner like me to miss. (I missed a lot; Cat interpreted.) Tapping someone with a branch or swinging an arm or a leg can mean "Follow me." The lesser-intensity version (as with us) is a wave of a hand, c'mon-style.

A human word can mean different things, depending on context. The English expression "Hey!" can be a friendly greeting or a hostile warning. We understand the user's intent because we understand context and tone. For chimpanzees, shaking a leafy sapling can mean come closer *or* move away. What is meant depends on context. Are we in a grooming context, or is the context sex, or general friendliness, or are you annoying me, or am I afraid of you, or have we just had a fight? When leaving, making a "come closer" scratch becomes an invitation to follow. Chimpanzee gesture types are each used for an average of three different goals. (In human toddlers, each gesture type gets used for an average of two goals.) But many gestures can mean essentially the same thing; there is substantial overlap. Chimpanzees use their sixty-plus-gesture repertoire to express only twenty or so meanings.

"Stop that" is often indicated by a slap on the ground, but about half a dozen other gestures mean the same thing. Why a choice? Cat explains, "You'd tell a high-ranking male to stop in a different way than you want to tell your mother or your kid. And if I am high-ranking, I want to reassure you before I come and embrace you; I understand it's a scary situation for you." This sounds to me like something we might call etiquette.

If a chimp's hand is outstretched, like a human about to shake hands with someone, it's usually for friendly contact. If you're a chimp and you extend your arm like that but your fingers are curled in, it's usually because you're greeting someone high-ranking and you're nervous about your vulnerable fingers. Males sometimes "show trust," Cat says, by doing a testicle tickle to one another. That's certainly trust — testicles are prime targets during truly violent fights; losers can come away without theirs.

A bare-teeth grin usually indicates nervousness. It's easily seen in these chimpanzees. It was formerly called a "fear grin." But that misnomer has fallen from use. Its message is: "Look, my teeth are closed; my jaws are not open to bite. I will not be attacking you; I am safe and I regard and approach you with peaceful intent." The human smile of greeting — the assurance that we intend friendliness, especially to a stranger — originated in this "bare-teeth grin" that communicates peaceful intent. It's a show of emphatic nonaggression. Immediately, automatically, the recipient of a smile feels more at ease, and tensions de-escalate. (We smile when meeting people with whom we intend to do business, and flight attendants smile so much not because the situation is happy fun for them but more because a smile signals friendly intent, which dissipates tension so that we can interact constructively.)

Chimps beg — for meat, for fruit, for sponges — with an open hand with the palm upturned. It's our beg, too; we likely inherited it from our common ancestor. While I am here in Budongo, a study going on in Europe and Uganda is finding that human toddlers use fifty-two distinct gestures, and chimpanzees use forty-six of those *same* gestures — an 88 percent overlap. Chimpanzee and bonobo gestures also overlap 90 percent. At least thirty-six specific gestures are shared among *all* of the non-human great apes.

This overlap suggests that all of us — all of the apes, including humans — inherited from an older common ancestor the ability to create, learn, and use these ancient packets of meaning, signals that have been flashed through Africa's forests for millions of years.

To really see how an ape thinks, you have to see communication *failing*. A good way to find failure is to watch an ape trying to communicate with a human. If a captive orang-utan, for instance, is trying to get a banana and the human offers a cucumber, the orang will switch to a totally different signal. But if the human is close to getting it — offering one banana from a bunch when the orang-utan wants the whole bunch — the orang will keep repeating the same cue. When someone is far from getting it, you switch your prompt; when their guesses are close to correct, you repeat and intensify the same cue. This is similar to how humans play charades. Apes — and quite a few other creatures — understand partial success and the difference between failure, partial success, and total success. Chimps are good at understanding whether they are getting their message across. They usually persist until they get the response they're looking for.

Humans are incessant talkers, so communicating without language seems a surprising proposition. But let's not underestimate the power of gestures. Opening our arms to offer a hug is not language. But it communicates a profound message.

We've seen that an arm raise means "Come closer." A scratch across the belly is an invitation to groom. An arm raise followed by a scratch across the belly means "Come closer; let's groom." Calls that convey "Where are you?," "Move away," "I see a predator," or "I've found food" might not be language in the human sense, but they form a rich shorthand that freights enough information to get the world's work done, generation to generation, for ages at a clip.

People say that you can't think without language. Other creatures show how it's done.

When a raven calls to their friends, "Food here," they show us their thoughts, in their language. Ravens' expressions may be spare, but the fewer the words, the more akin to poetry.

"Many people don't consider it language unless the ordering of the parts can change the meaning," Cat acknowledges. ("I have spotted a dog" and "I have a spotted dog" mean different things.) But she parries this objection: "That's important in how humans communicate. It's less important if your goal is to understand how chimpanzees communicate."

Cat would be the first to tell us that there's still a lot we don't understand about how chimpanzees use their vocalizations and gestures, and what meanings they convey. But back in the 1960s, our ignorance was nearly absolute. In that era, early attempts to get at the language capacities of other minds entailed plucking those minds bodily from their natural social context and delivering them as captives to us. Slowly, a bit clumsily, researchers began to tap into the mental workings of apes, dolphins, and parrots, with language training, signing, and language-related symbols. The quest was not to understand their lives, and not quite to learn how *they* communicate, but to try to get them into conversations with *us* — often in English. Because chimps cannot make human-like vocal sounds, they "failed" to learn English. We really can't make their sounds, either, but no human considered our inability a "failure." Initially the idea was to "test" them, and the tests were limited to, essentially, "Let's find out just how smart these things are."

In 1967, Allen and Beatrix Gardner, at the University of Nevada, managed to acquire a baby wild-born chimpanzee, whom they named Washoe. Aware of failures to teach chimps English, the Gardners thought a chimpanzee might be able to learn sign language. At least, they would explore her communication capacity without running aground on the limits of her vocal cords. Mindful of her needs for socialization and group identity, the Gardners raised Washoe in their home as one might a human child.

Washoe did learn to sign, developing a 350-word sign vocabulary. She acquired various signs by watching people signing, without being taught. Watching elders is, of course, how chimpanzees usually learn everything they need to know.

One day, upon seeing a swan, Washoe signed "water" and "bird." Humans combine words similarly and, in fact, consider swans to be among the "waterbirds." Washoe named watermelon "candy fruit." She would sometimes climb a certain tree, from which she could see arriving cars, and, by signing, would inform her earthbound humans of the identities of the coming visitors. Washoe was taught the sign "open" for opening doors and containers; she spontaneously generalized "open" to water faucets, as we tend to do. She was taught the sign for "dirty" in relation to feces and unclean things. She then used "dirty" as an adjective before the name of someone or something that displeased her. Thus Washoe created insults by using simple grammar. Another chimp, Lucy, also used "dirty" in this way. When Washoe broke up a fight between her adopted son and another juvenile male, she signed, "Go." She'd previously used that word only as a request, but she had now turned it into a command.

When Washoe was in her early teens and was living with a group of sign-trained chimpanzees, she was given an adoptive son, Loulis. By plan, Loulis received no instruction from humans, and humans never signed within sight of him. Washoe not only taught Loulis signing, she — astonishingly — taught him by "molding," shaping the hands of her pupil into the sign while putting them through the associated movement. Once when a candy bar was imminent, Washoe excitedly signed "food" repeatedly while making many food grunts. Loulis, sitting next to her, watched. Washoe stopped signing, molded Loulis's hand into the "food" configuration, and put it through the sign's movement several times. Loulis eventually acquired about seventy signs — with no human instruction. This represented a previously unknown cultural capacity. It was groundbreaking.

In the 1980s, the researcher Sue Savage-Rumbaugh began working extensively with several bonobos, especially the bonobo Kanzi. Kanzi understands three thousand spoken human words, understands syntax in sentences like "put the ball in the refrigerator," and can use symbols to ask for marshmallows and a lighter — and then make a fire and toast them. (You can find some impressive Kanzi videos online.) Savage-Rumbaugh described an experiment where Kanzi and his sister Panbanisha were in separate rooms where they could hear, but not see, each other. They had both become proficient at using keyboard symbols called lexigrams that stand for words or objects but are *not pictures* of the objects and bear no visual relation to them. (For instance, the lexigram for "automobile" is two curved red lines.) Savage-Rumbaugh explained to Kanzi that he would be getting some yogurt. She asked him to vocally tell Panbanisha. Kanzi vocalized. Panbanisha then selected the lexigram for "yogurt" on her keyboard and vocalized back. Some observers believe that the high, squeaky vocalizations of bonobos are language-like, densely informative communication at a speed currently beyond human comprehension. Untangling this would require detailed study and analysis, but for now these possibilities remain tantalizingly unexplored.

If you watch chimpanzees and bonobos using computer screens, you realize that they can indeed mentally process concepts and respond at speeds that humans cannot even follow. Search online for: "chimp vs. human memory test." You'll see part of a BBC documentary. In a setup where the numerals 1 through 9 appear in random positions on a computer screen, the task is to press them in the correct sequence. The catch: as soon as you hit 1, they're all replaced by blank squares. Now you have to hit the blanks that hide the numerals, in the correct sequence. Humans stare at the numbers, trying to memorize their positions before they hit 1. They get it right about once out of thirty tries. A chimpanzee just glances at the numerals, hits 1 and then the remaining eight blank squares faster than your eye can follow, and gets the sequence right 80 percent of the time. This shows that there are differences in our mental capabilities, and that in some ways, at some things, chimps are mentally better and much faster than humans.

Signing opened a window to a previously unseen view of the ability of apes to imagine, communicate, conceptualize, generalize, share, and spread new culture. But by the late 1970s, the field had begun to contract and decline. Funding started drying up, the first generation of language-enculturated apes began dying off, and the next generation of students were more interested in how human brains work than in the lives of other minds.

And some people, deeply moved by what those earlier studies had found about the minds of other kinds, began campaigning to end all captive research on the most communicative species. Granted, much of what apes and dolphins endured in research and show business was cruel. But the baby who was learning to talk got thrown out with the bathwater of concern over welfare and drowned in regulation. Is that good or bad? It feels, unfortunately, necessary.

Irene Pepperberg, who pioneered communication with the African grey parrot Alex, laments, "As a consequence, we are missing many opportunities . . . to deduce the precursors to modern human languages [and] what types of brains our ancestors might have been developing."

If "precursors to modern human languages" is *why* we were interested in communicating with apes and parrots, then we weren't really interested in them. We were interested — as usual — only in ourselves. We weren't trying to communicate with or understand or appreciate living beings. If it's always about us, then ending these studies won't cause us to miss out on anything; we were, all along, missing everything. It reminds me of the old joke about a man on a first date saying, "Well,

I've spoken quite enough about *myself.* What do *you* think about me?"

Pepperberg says the heyday of the research was a "Doctor Dolittle moment." The researchers had made a breakthrough toward one of the great dreams of humankind: they'd begun talking to the animals. Better, perhaps, if we'd quieted our chatter and tried *listening.*

When we stop viewing the apes and parrots and dolphins as "models" of primitive human origins — when we remove our human blinders — we can see that other beings on Earth take an interest in their lives and understand something about who they are, where they are, who they're with, and what they're doing. In their shared experience, they sow and reap their own cultures.

Who we are voyaging with on this lonely living planet is quite possibly the only meaningful line of inquiry. I knew that Cat and I would get along when she said, "I haven't come to work in a remote Ugandan forest because I am interested in 'clues to human evolution' or 'how we became toolmakers.' I'm here because I'm interested in chimps."

PEACE:
SIX

This afternoon the chimpanzees have elected to linger at their source of water for the uncommonly long span of more than two hours. One of the chimps still has a serious-sounding wet cough. In broad daylight, he bends branches into a nest and lies down in it.

"He must be feeling miserable," Cat comments. In the two-decade span of this research about the chimpanzees of Budongo Forest, this year has been the worst one for them catching life-threatening colds. When people enter the forest to cut wood or set snares, they sometimes leave plastic cups at water holes. A curious chimp might pick one up. That's one way a person can pass on an infectious strain that chimpanzees have no experience dealing with. (Chimpanzees elsewhere have also died from human-source infections, including, in fact, the rhinovirus — the common cold.)

Ten days ago, the coughing cold claimed

Ketie's two-year-old daughter Karyo. Ketie is still carrying her daughter's body.

"I think she understands the baby's dead," Cat says. "But —"

It's a pitiful sight. In the moist morning air, the smell of Ketie's putrefying infant trails her like exhaust.

Three small chimps, strangers and apparently newly orphaned, descend the slope: a juvenile boy and a girl of perhaps six years carrying a young baby. It's likely that these lost children are a brother and sister and their very young sibling. The baby is about a year and a half old — only half the age where survival without milk might be possible. The sister clutches the baby protectively to her body. The little one, too weak to cling, cries continually. Whatever happened to their mother occurred within the last couple of days.

Cat has already thought through the possible remedies. There are none. "The downsides of intervening," she says, "would be worse." Capture would be disruptive and terrifying to all the chimps. And a baby chimp "saved" in captivity would be captive forever. Such a chimp could not be rewilded to live a normal chimpanzee life.

A baby under five who loses their mother is usually doomed. Any other nursing mother can't produce enough milk for an additional infant. Even if weaned (earliest, around age three) before their mother dies, their psychological and social needs for their mother are so profound that young orphans sometimes lose the will to live and let themselves die. "They simply can't make it without their mothers," writes Christophe Boesch.

That's not *always* true. What is true is: they need *someone*. A companion in life. Sheer emotional support. Newly orphaned siblings usually stick together; the older one grows up fast and acts like a child head of household. Ten-year-old Spini and her little four-year-old brother, Soldati, lost their thirty-year-old mother, who disappeared a few months ago. But their luck is holding. They have been playing with Lotty, Liz, and Monika. Cat is pleased, adding, "For little Soldati, it was iffy for a while."

Happily, a weaned orphan with no sibling still stands a good chance of being adopted. Adoption is a months-long process of gradual bonding. In some cases, a close female friend of the deceased mother will adopt an orphan. Adopters carry orphans and bridge gaps while climbing in trees, either with their bodies or by bending a branch. While traveling, they'll wait for their adoptees, intervene in conflicts, and share food.

Adult males normally don't provide *any* parental care to their own offspring (they almost certainly have no inkling of paternity). So it's surprising that adult males not only adopt but show many "maternal" behaviors, even letting adopted kids ride on their backs. If alpha, they'll share meat.

Barring adoption, a juvenile orphan surviving without a guardian can become permanently stunted in size, low in status, forever marked by the loss of her mother early in life. An orphan who is ten might look six. Ten-year-old Lillo is indeed small, having scrambled to make it without a mother, quickly learning to forage, travel, and groom his way into relationships like an adult, well beyond his years.

Our planet spins a weave of tragedies. Life is bearable only because the warping maladies sometimes come with wefts of little triumphs.

Leaving the water holes, we walk a steady incline on a long trail for about seven minutes. Though the trail runs along the slope of a hillside, its well-worn track has been tramped level; chimpanzees have walked here for long generations. I hear a sudden grunt from overhead. We abruptly stop. I look up but see nothing.

Kizza sees Ketie, clutching her baby's corpse. "Chimps whose infants die soon after birth," Cat says, "seem to bounce back more quickly than mothers who've had time to form the mother-child bond." Ketie and her baby had been together day and night for two years.

In 2018, the orca whale known as Tahlequah, or J35 of the Pacific Northwest's J Pod, pushed her dead infant along the sea surface for seventeen days, a distance of a thousand miles. Ken Balcomb, who has studied these whales for half a century, called it "a very tragic tour of grief." Tahlequah's baby had lived for half an hour after birth; Tahlequah experienced her child alive and began the intense mother-child bonding that, in the exceptionally close-knit culture of those orcas, normally lasts for the rest of the mother's life. Because these whales' newborns continue dying from a combined dearth of fish and abundance of toxic chemicals, the *New York Times* — in a rare official obituary for a non-human — said that Tahlequah's funereal thousand-mile procession "felt like more than mourning. It felt like an accusation." An accusation because the ongoing destruction of wild salmon — by overfishing, fish farms that generate fatal parasites, dams, logging, and polluted streams — is why these whales are starving and their babies are dying. Chimps aren't the only beings who have trouble letting go.

Ketie has been carrying her dead infant for ten days now; one West African chimp carried her dead baby for twenty-seven days.

When people ask whether "animals" have "a concept" of death, I ask whether humans

491

have "a concept" of death. Human concepts about death are many and varied. Some people believe that when you die, you cease to exist. Many believe that in death they'll be reunited with the loved ones of their Earthly existence. Most people hold faith in some eternal life, ranging from a wheel of karmic rebirths to heaven, or damnation, and so on. The Incas believed their emperor immortal and treated his mummy as if it were undead. Miguel de Estete, who accompanied Pizarro, described dead Inca emperors "seated on thrones and surrounded by pages and women with flywhisks in their hands, who ministered to them with as much respect as if they had been alive." The mummies spoke through mediums, giving advice and directives. They retained all their wealth: none of it could be passed on; palaces could not be reoccupied. This created enormous drag on the Incan economy and seeded divisive politics. So that's one extreme. Humans *usually* understand a difference between alive and dead, but humans do not have "a concept" of death. They have many.

"Chimps understand death at some level," Cat points out. "They kill monkeys. Sometimes they kill other chimps. So they have to understand death."

Indeed, they do. Sometimes they cause it; sometimes they recognize a fatal accident for what it is. When a chimp died in a fall from a high tree, other chimps gathered, stared with expressions sometimes referred to as fear-faces, and embraced one another.

One closely related issue is whether non-human beings like Ketie and the whale Tahlequah "grieve." Regarding human grief, Cat clarifies the range of reactions: "You and I have a Western concept of what grief looks like." Even grief is cultural. "More than once, one of our field assistants has come to work, we've worked together all day, and I find out later that the night before, their baby died." Infant death happens a lot for people here. Being unable to function for days or weeks because we're grief-stricken — "that's a luxury *we* have." Here in rural Uganda, the outward response is different from the out-pouring of emotion shown in Western culture. "But they *have* lost a loved one," Cat empha-sizes. In many cultures, people don't name their infants before they're three or four years old. They think it's much more of a loss if a baby with a name dies. "When a new baby chimp appears," Cat adds, "*our* impulse is 'Let's name it.' The field assistants are very reluctant to name them before they're about two years old. To them, the act of naming changes the relationship."

When a chimpanzee loses a mother or an older child, they grieve. "Same things you'd see in a person," Cat observes. "Time alone. Sitting quietly at the edge of a group. Not getting involved in social goings-on. Sitting staring at the ground. They eat less. Lose weight. They might sleep for a whole day or two, acting listless and depressed." Scientists who've considered the question have written, "Chimpanzees' awareness of death has been underestimated." But what in the non-human world has *not* been underestimated?

In 2001, on the day an adult Sonso female, Ruda, was dying, she was surrounded by two dozen chimpanzees. The notes of the day include several mentions of the chimps making "strange calls." The alpha male at the time, Duane, "looked very scared" and kept a spooked distance. Others came close, then ran away. One male struck Ruda, "to see if she could get up . . . in vain." At one point, all the chimps left except for Bob and Rachel — Ruda's children. Bob was eleven at the time; Rachel was a little girl of just four. As they gazed at their dying mother, the notes continue, "Bob screamed loudly, which I termed 'crying.' " Then for twenty minutes they both uttered peculiar-sounding screams. "We really felt so sad when they started screaming," the notes say. Ruda died during the night. After that, young Bob stayed close to his sister. Both survived. Sometime later, a researcher saw them and remarked, "Bob & Rachel — the small and tiny orphans — wandering alone. But fine."

■ ■ ■ ■

The sight of Ketie carrying her dead baby has become particularly pitiful. The corpse has dehydrated, the legs shrunken around the bones, the feet like paddles at the ends of sticks. Perhaps her hormonally driven maternal urges are simply too strong. Perhaps her understanding of death is incomplete. Or perhaps Ketie *is* grieving, and won't let go. We may be reluctant to say that grieving humans are hormonally driven or have an incomplete understanding of death, but that is often true of human grief. We may be reluctant to conclude that Ketie is grieving, but that is as likely true.

A young female chimp will sometimes carry a short, thick chunk of wood against her body for a few hours. Usually they hold and treat these objects gently. They often hug and sometimes even groom them. Researchers call these objects "log dolls." Young chimps sometimes carry such pieces of wood — or sometimes stones — up into trees and place them next to themselves, or even build nests for them and position the object in the nest. (Orangutans sometimes sleep with leaf-bundle "dolls.") Wrote noted primatologist Richard Wrangham, "They're treating them like a baby." The behavior is "difficult to explain unless they're conceiving of these like a baby." After chimps have a real baby, they are never seen with such dolls. The chimps, added Wrangham, "seem to be imagining something about ordinary objects, something about a relationship with another individual." Most important, I think, is that by investing emotion and a concept into these mere objects, they make them *symbolic;* they create meaning.

I've begun to look forward to our long walks homeward through Budongo Forest each evening, for the long conversations and recaps of our day's observations. Today, in the lengthening afternoon shadows, Cat tells me that there are two sides to chimpanzee emotional drama.

PEACE:
SEVEN

On the walk back to camp, Cat says that, yes, the hooting and screaming and hysteria are designed to get your attention. They do just that. We see chimpanzees' contagious excitement and contagious fear as easily as we see the bright side of the moon. And because we do, the other side of chimpanzee emotionality, their sympathy, gets overshadowed. The deeper nature of chimpanzees also includes tender empathic concern for others and brave altruism. Always there, seldom visible. It sometimes comes to light by unusual means.

Nearly a century ago, the Russian researcher Nadia Ladygina-Kohts wrote, "If I pretend to be crying, close my eyes and weep, Joni immediately stops his play or any other activities, quickly runs over to me, all excited, and from the most remote places in the house." For several years in the early 1900s, Ladygina-Kohts observed — in detail far, far ahead of her time — a young male chimpanzee she'd named Joni, who lived with her.

"Looking at my face, he tenderly takes my chin in his palm, lightly touches my face with his finger, as though trying to understand what is happening. . . . The more sorrowful and disconsolate my crying, the warmer his sympathy. He carefully places his hand on my head, extends his lips toward my face and looks into my eyes compassionately and attentively. Then, he . . . touches my face or my hands with his extended lips, and slightly pinches my skin (as if kissing); sometimes, he touches me with his open mouth or his tongue."

Human two-year-olds behave almost identically. Dogs respond to distress similarly to human toddlers, pawing, tugging, licking, and seemingly wanting to console. Elephants, whose emotional bonds to one another are profound, often assist and soothe another elephant in distress. In experiments, rats will often open an experimental door to save a rat in danger of drowning — especially if they themselves have experienced a nasty immersion in water — even if it means forsaking a chocolate treat. But if the pool is dry, meaning the other rat isn't in trouble, they perceive no need to open the door.

Empathy is the ability of one's mind to match moods with one's companions. When we do not exactly feel the same but we understand something about the experience of another being, we can call that form of empathy "sympathy." To act from sympathy to try to ease a troubled other by soothing, consoling, or aiding is to show "compassion." Feeling with another, feeling for another, and acting to help are, I believe, three levels of empathy. When we open our eyes to other species we see: empathy is widespread. Humans don't have a monopoly on empathy. We have, in fact, a long way to go to perfect it. (Intentional cruelty and torture require a mind equipped with sufficient empathy to understand that another is suffering.) But in our finer moments, human compassion is the best thing about us.

Sometimes acting out of compassion for another's benefit brings chimpanzees into danger. The word for that: altruism. When a companion is unaware of a danger, a chimp is more likely to sound an alarm than if it's obvious that the companion is also aware of the danger. Once, a chimpanzee was seen holding back a companion who seemed too interested in a potentially dangerous snake. In two separate incidents in zoos, a mother and a male chimpanzee drowned while trying to rescue babies who had fallen into a moat surrounding their enclosure. Washoe, that first chimp taught to communicate through signs, ran across two electroshock wires and then risked her life to successfully grab and rescue a female chimp who'd fallen into water and was screaming and thrashing — a female Washoe had first met just a few hours earlier. Zoo chimps sometimes bring food or even mouthfuls of water to elderly group-mates.

Altruism in zoos is one thing. Then there's real life. In Taï Forest, when four males penetrated deep into the neighboring community on an offensive venture, researcher Christophe Boesch saw them rush and attack a female with an infant. While the intruders beat her on the back, "she crouched on the ground protecting her baby." Suddenly a second female from her community appeared and attacked the aggressors. This allowed the first to escape. But the female who had come to the rescue was not so lucky. The males quickly subdued her, biting and beating her. Luckily for her, males of her own community appeared and routed her attackers. Had she been human, her efforts on behalf of the first female would be called heroic. What difference should her species make?

In another case of altruism, a male named Porthos, with an adopted baby daughter riding on his back, rushed toward distress screams from a female named Bamu. Five males of an adjacent community had captured her. Bamu had years earlier lost an arm; she was helpless against her attackers. With his adopted baby clinging, Porthos charged the five males so fiercely that he saved Bamu. Porthos had certainly risked his life; that very

500

day those same five intruders killed a male of Porthos and Bamu's community.

And if a leopard has grabbed someone? "Chimpanzees do not hesitate," Boesch writes, to rush toward alarm calls of individuals in mortal danger. Leopards have vise-grip jaws and eighteen slashing box-cutter claws. Help must arrive within seconds. One day, a female chimpanzee was attacked while rescuing her small son from a leopard. An adult male chimp rushed immediately to *her* rescue — and *he* was in turn attacked. Over twenty-five years of observation, Boesch has come to see such heroic altruism *not* as the exception, he says, but as typical of chimpanzees.

The altruistic capacity to risk personal safety for the safety of other group members can form a net of cooperation that puts group life, group identity — and, therefore, culture — on another level. After the rescuing male was injured by that leopard, all the females and some of the males in the group cared for him for hours, wiping away blood and cleaning dirt from his wounds. The cleaning and care were not unique to that one attack. "All the injured victims of leopard attacks were attended by other members of the group," Boesch notes. "Caring included carefully licking the wounds and removing dirt from within the injuries. . . . An injured individual receives support for days and, for wounds on the back, head, or neck, tending is extremely important." (Unfortunately for the male who came to the rescue of the female, one of the leopard's scimitar claws had perforated his right lung, and six weeks later he died of infection.)

Here in Budongo Forest, when Zig appeared limping badly and with a deep wound on his leg, Pascal groomed it and sucked on the wound for about three minutes. The wound healed quickly, and Zig soon resumed full motion and activity.

Beyond the bluster of status seeking lies something much deeper in the chimpanzees' collective social soul. Empathic concern and altruism confer enormous survival advantages. If they didn't, chimps wouldn't live in groups. Nor would humans. And if humans didn't, cooperation would never have spread to every corner of Earth.

To understand whether your actions are helping or hurting — and when you are being helped or harmed — you must be capable of sympathy and compassion. Again, humans didn't invent that. Closely related is the ability to know that another would want what you want, and this leads to a sense of fairness.

Our dogs keep an eye on each other, and if someone gets something good, they all want to get something at least as good. Fair is fair. Sometimes when one of our hens doesn't feel like commuting back to the coop, Chula finds an errant egg under the porch. Our deal is: no coop-raiding allowed, but she gets to eat any wayward eggs she finds. I can ask Chula to give me an egg — and she will, unbroken — but she knows that if she's just found it outside, I'll immediately give it back. Finders keepers; that's only fair. So, other beings sometimes perceive fairness. But is an insistence that fairness should flow *to others* — generosity — exclusively human?

In experiments, each of two chimpanzees got either a sweet grape or a less desired carrot. If only one got a grape, the carrot holder sometimes turned grumpy, tossed their carrot, and went on strike. Not surprising. (Monkeys behave similarly.) "But no one had anticipated," noted coresearcher Frans de Waal, that grape-receiving chimps, too, sometimes refused their own reward if their partner got only a carrot. "This comes much closer to the human sense of fairness," wrote de Waal.

When a token of one color rewards the chimp who chooses it and selecting another color rewards both the chooser and another chimp in an adjacent enclosure, 70 to 90 percent of the time the chooser selects the color that rewards both chimps — about the same percentage as with seven-year-old humans. Chimps also pass tools to comrades in adjacent cages so that they can access treats beyond their reach. The bonobo Pan-banisha sometimes refused an experimenter's goodies and gestured to her watching family members until they, too, got treats.

Frans de Waal says, "We are fair not because we love each other or are so nice but because we need to keep cooperation flowing . . . retaining everyone on the team."

But very often we *do* love each other, and we *are* nice. We must not dismiss it; we must account for it. We *do* love and we *are* nice *because* we need to keep cooperation flowing and our team functioning. We need one another. And sometimes, like chimpanzees rushing to the rescue, we simply understand when someone needs us.

When Patricia and I brought home our new seven-month-old Aussie shepherd pup, Cady, our six- and seven-year-old pooches, Chula and Jude, hated the intrusion. Chula's response ranged from giving us "What the —" looks to occasionally being aggressive when the pup tried to play. (It didn't help that, having spent her first months in a city apartment and having never been socialized — or house-broken — Cady's play signaling was off.) Jude kept going to the door simply to get away from Cady's compulsive barking. (I must say I empathized.) Neither older dog wanted anything to do with her. (That, too, was understandable; my wife and I were asking ourselves why we had ever agreed to take such a challenging creature into our peaceable kingdom. But we knew the answer: we'd been asked to help and we'd said yes.)

One morning after Cady had been with us for a month, we were on a beach where we often let them run. We were almost back to our car when Cady decided to turn and run after a jogger's dog several hundred yards down the beach. If you want to get a young dog in the habit of following, you have to let them realize that you won't wait for them if they hightail it on you, so I kept walking to the car. Chula and Jude, though, didn't share my training concepts. They stopped about sixty paces from me, looking toward the receding puppy and then at me. I called them. Chula came, slowly. But Jude refused. He sat. I called again. He *lay down,* butt to my direction, facing the distant bouncing, barking dot. I kept walking toward the car, periodically glancing over my shoulder. Cady was now running back to us fast enough to complete a three-minute mile. When she reached the lying-down Jude, he leapt up and followed her, watchfully bringing up the rear of the entourage. Now we *all* understood that we belonged together. Jude's unilaterally enacted "no pup left behind" policy surprised me. Outwardly he's the mellowest of all of us (we call him "the poet" because he often seems on a faraway plane of being). But every now and then, his actions reveal that he is paying closer attention and caring more than you would guess. Jude seemed to be implying, "In *our* group, this is how *we* do things" — the most fundamental concept in culture. For non-human beings, actions speak louder than no words.

I wonder whether Cady appreciated Jude's thoughtfulness. I know I did. In acting on the sentiment "I won't leave unless we all leave," Jude was being both tribal and fair-minded.

A sense of fairness is more widespread than people might expect. But perhaps that's because our own expectation of fairness has become so broken. In hunter-gatherer human groups, no ethic appears more fundamental than the expectation of sharing. Yet in our own lives, nearly all social problems are symptoms of inequalities, rigidly enforced. Humans have a great sense of fairness, but in *practice* our fairness isn't great.

"If you now ask me if there is any difference between the human sense of fairness and that of chimpanzees," de Waal writes, "I really don't know anymore."

The same mental ledger that allows us to reciprocate our gratitude also enables us to plan revenge for poor treatment. Your dog knows what to expect from you based on your history together; an elephant can wait years for just the right moment to return the favor to a sadistic keeper or to embrace an old friend not seen in decades. Fair is fair. But to connect acts of the past to the present and the future requires a remembrance of things past, and a sense of time. Many people have believed that only humans are capable of these things.

But go back in evolutionary time. Further. Much further. Even some bacteria — one cell, no nervous system — somehow use a time sense to guide their action. And though it seems unlikely that bacteria experience *feeling,* their complexity is mind-blowing. For a bacterial cell to approach the nutritious and avoid the noxious, writes biological philosopher Peter Godfrey-Smith, "one mechanism registers what conditions are like right now, and another records how things were a few moments ago. The bacterium will swim in a straight line as long as the chemicals it senses seem *better* now than those it sensed a moment ago. If not, it's preferable to change course."

In ways that required adding hundreds of millions of years of increasing complexity and sophistication to the decision-making systems in bacteria, chimps and humans now know who they are, who they've been with, and who has done what to them.

Various apes and other primates, as well as those crows we've mentioned and, of course, our dogs, have certain expectations about social norms and fair distribution of food and will protest if they are cheated. In experi-

507

ments, chimpanzees who've had their food stolen sometimes "punish" the thief by pulling a rope to move their food reward out of reach. In a test devised for a human sense of fairness called the "Ultimatum Game," chimpanzees protested selfish partners by spitting water and hitting the cage. Human children in the same study protested by saying things like "You got more!"

Why be fair? Why not get as much for oneself as we can get? And why not cheat — why not take what might have been intended for or belong to someone else?

This is a question that only modern people can afford to ask. We live where it's actually possible to get away with cheating. In more natural communities where everyone knows who's who and keeps their mental tally, cheating doesn't work well. Sharing and caring do. In fact, a leading hypothesis for the evolution of intelligence itself is that individuals in social groups need to keep track of others' actions, of their history, of the benefits they promise and the risks they pose. Consequently, intelligence evolved to fill the need for a *social brain* capable of planning, plotting, paying, punishing, protecting, seducing, sympathizing, loving —. To know what, you must know who.

So beware of the spreading anonymity.

PEACE:
EIGHT

As usual, we reached camp just after sun-down, ate our rice and lentils well after dark, went to our dreams after setting our alarms for well before first light, woke with the hyraxes' screams, drank our bad instant coffee, and marched our headlamp-assisted trek through the forest to arrive in the chimps' Waibira territory by sunrise.

Now it's a new day, and Monika and nineteen-year-old Alf have a thing going. Monika is old enough to act sexual, but probably not yet capable of conception or successful motherhood. Yet right now she is sporting the characteristically large swelling of her bottom that advertises her estrus to one and all.

Estrus, a time around ovulation when a female is sexually motivated, occurs periodically in most mammals, perhaps all. Also in most mammals, ovulation is advertised with physical and chemical cues. Female chimps start getting small estrus swellings around age ten. It takes a few years for them to begin having full estrus swellings and to become fertile. Between periods of estrus, female chimps are not sexually active. All great apes have estrus cycles. Adults usually get pregnant during each cycle, and pregnancy and nursing pause the cycles for several years. If fertilization does *not* occur, great apes (along with many primates, including humans) menstruate, meaning that the body flushes rather than reabsorbs the lining of the uterus, the endometrium.

As for a female chimpanzee's estrus swelling, "subtle" is not the word. It's a *very* pronounced swelling that becomes increasingly turgid, until reaching the size of a melon at its peak. Estrus spans ten days to two weeks, after which the swelling shrinks away. It would be as if a human woman, with every menstrual cycle, grew large breasts that disappeared between cycles. Female chimps in captivity can go through five or six cycles a year, but free-living adult females are often pregnant or nursing; they typically spend over 80 percent of their life caring for dependent children.

Females in estrus often actively seek sexual contact. "We had one," Cat recalls, "who would back a ranking male out at the end of a branch, as if 'You're not getting past me until I get what I'm after.' " But because chimpanzees live in groups wherein males are in competition and females have their pick, males often invite females for sex, by raising an arm or shaking a branch or clipping leaves with their teeth in a particular way. These are coded messages. Everyone understands them. (In the bonobo version of leaf clipping, they tear and drop pieces of a leaf, tear and drop, as in "She loves me, she loves me not." Except that in bonobos, she always loves him. "Or her," Cat adds with a smile. "Or sometimes — them.")

It doesn't take much of an invitation, usually, for a female chimp in peak estrus to come right over, turn round, and back onto him. On the waxing and waning sides of her peak, she might copulate twenty times in a day with about ten different males of lower status. That's not to say she's not choosy. Females usually avoid mating with males they don't want to mate with. Further, in one study females interrupted more than 90 percent of the matings they were involved in. At peak, around ovulation, she clearly prefers high-status males — if she can get them.

At fourteen years of age, Monika isn't yet mature enough to be of serious interest to high-ranking males. In humans, young females often prefer older, more proven males; older males often prefer young females. In chimpanzees, females *and* males prefer older, more proven individuals.

Females typically begin breeding at fifteen

or sixteen. They often lose their first baby. It's not just inexperience, though a first-time mother is more likely to make mistakes. Usually, they've recently moved into a new community, a new place where they don't fully know the territory or fully participate in the social dynamics. The stress of moving gets compounded by not having strong allies. Their babies are more vulnerable. Monika was born in the adjacent Sonso community. She immigrated here to Waibira three years ago.

Alf is neither adult nor high-ranking, but he's interested. Alf shakes a sapling. Monika doesn't respond. He shakes it again, more emphatically. She moves closer. Striking a bit of a deal, perhaps, Monika solicits grooming. When Alf again leans back with an arm raised and legs man-spread, advertising his pink readiness, she spins round and backs directly onto him. They must like the feeling, but the act is so brief it seems perfunctory. They seem to savor grooming far more than mating.

Macallan very deliberately tears a leaf; it makes a distinct sound. Monika moves toward Macallan, spins, and backs into him with startling speed and precision. Even Cat thinks that was very fast. Chimps show no apparent desire to prolong the act, none of the recreational enjoyment seen with bonobos, let alone the tenderly attentive artfulness or high-arousal eroticism that sometimes flows in humans.

Chimpanzees do not seem to show the affection seen in creatures with pair-bonding. A few other primates, such as the *Callicebus* (titi) and *Aotus* (owl) monkeys, as well as gibbons, form pair-bonds. Otters, gray wolves, prairie voles (surprisingly), and many birds — ravens, city pigeons, owls, especially albatrosses and parrots — share the sort of affection that helps bond breeding pairs, who will share parenting responsibilities and remain mated for years. Humans, of course, often pair-bond. Chimps, though, don't have that pair kind of bonding; chimps have friends. With benefits.

Monika walks about fifty yards. She next copulates with Ardbeg.

It appears to be a lucky day around here — if you're not Masariki.

Masariki, just sitting there, has an erection. He is a teenager — which means low status. He's an orphan — even lower status. He desires. He is not desired.

After half an hour of their couplings and our voyeurism, they move. A few at a time, they head off. Somehow they all seem to know where they're going next.

They melt into a part of the forest so thickly tangled it's as if they want to filter us out of their lives for a while.

They succeed.

■ ■ ■ ■

A couple of hours later, I can barely hear some very distant chimp calls. Kizza again astonishes me by identifying the authors of these faint calls as Lafroig and Douglas.

The chimps' destination is an enormous tree. It's *Strychnos mitis,* in the genus of plants whose stems and leaves defend against plant eaters with the deadly poisons used for making strychnine and curare. But the tree needs a way to get its seeds out and around, so the fruits are nontoxic. The chimps hoist themselves aloft like dark flags and get busy. Each yellow, cherry-tomato-sized fruit is little more than a big pit whose nutritious but stingy covering is skin-thin. But the tree is festooned, and the chimps suck them by the hundreds, spitting pits, whose free fall through the leaves creates a soothing patter like light rain.

For this while, there's at least sufficient peace for them to eat and socialize. Different parties within earshot send greetings, news of who and of where. Our party pauses to participate in the call-and-response; then the patter of pits resumes.

All of the Waibira females who happen to be in estrus are up there overhead. "So," says Cat, "you've got good food, you've got sex — why come down?"

There's agitation in the treetops.

Lotty is up there. As a mature, high-status mother, Lotty ranks as one of the most desirable females. And with her gibbous estrual swelling peaking at its most taut and turgid, her tumescence is sparking commotion and lofty erections.

"The world will revolve around her for the next day or two," says Cat. "We'll see what Lotty does."

Lotty's swelling is dark all around its base; its terminal surface is a pink pad. In experiments, chimps shown photos of two faces and one behind easily match the face with the behind — as long as it's someone they know. For knuckle-walking apes and four-legged creatures (think of dogs sniffing other dogs), the main signal areas of identity and interest — faces and behinds — are on the same level. But if an ape evolves to walk upright, this introduces a signaling design flaw. What to do?

Several researchers have observed that walking upright on two legs made it beneficial to move human signaling up, and around to the front. Scientists thinking about human topography have suggested that "breasts evolved to resemble the bottom" while being more visible to those of us standing on two feet. "Also," write the researchers, "humans, especially females, developed reddened and thicker lips and fattier faces compared to chimpanzees." In humans, these distinctive signs are often further accentuated with makeup. "Thus," the researchers continue, "human faces share important features with the ancient primate behind . . . two hairless, symmetrical and attractive body parts, which might have attuned the human brain to process faces, and the human face to become more behind-like." The researchers conclude, "Thus, a behind inversion effect seems plausible. But on the other hand, there is not much literature on how humans process behinds, and the behinds that we see in our daily life are usually covered."

Sounds strange and amusing, but — those scientists might be on to something. Chimpanzees' mammary glands never swell into visible breasts, not even during nursing. Chimps never mate face-to-face. Bonobos' mammaries do enlarge to visible breasts during nursing, which for an adult is much of her life. And bonobos often mate face-to-face. Bonobos have lips colored a pinkish hue that's intermediate between chimpanzees' lip color and humans'.

When Lotty comes down, everyone follows her. Gerald and Macallan, Masariki, Daudi —. In fact, we are all in Lotty's retinue today. We suddenly notice that Ketie is no longer carrying her dead infant. We wonder, Did she just drop it? Or did she set it down, consider it, and deliberately decide to walk away?

Lotty pauses to rest. She is surrounded by a small circle of adolescent males.

Ben, the alpha, and Talisker, the senior statesman, sweep in. Fiddich bolts for cover, and they chase him. Thirteen-year-old Daudi nervily exploits the confusion; he mounts Lotty. But he doesn't achieve actual copulation.

As a high-status female, Lotty has the prerogative to select the finest sex — in her opinion — that Waibira has to offer. So why would she be fooling around with a juvenile like Daudi?

"I don't think it's her ovulation day," Cat surmises. "If it was, Ben and Talisker would be constantly guarding her. And Lotty would be choosier. Right now she's acting more horny than choosy."

Madame will be ovulating in the next day or two. Although the most turgid appearance of her swelling will span about four days of her twelve-day swelling cycle, she'll be fertile for only a day or two at ovulation. When she does ovulate, all the adult males will know that it's her time of times. Researchers detect ovulation by collecting urine; male chimps' brains come equipped with their own chemical labs, and they sample by sniffing, testing, and tasting the female's relevant end.

517

Because apes stand a good chance of getting pregnant and then not cycling again for another few years while they're nursing their baby, females and males must both respond to their brief window of fertility. During those hours, they need to be *very* interested in sex.

I doubt that chimps understand what makes babies. They don't need to; male chimps provide no parental care. I also doubt that females know that a baby has anything to do with something they did for a few moments eight months earlier. Their interest in sex seems motivated, for them as for us, by pleasure's sufficient reward. As for outcomes and the big picture, it only matters that they have sex with viable partners on the right day. A highly advertised ovulation is an excellent way of getting males and females together at precisely the right time.

Ben and Talisker want to move Lotty away from the lesser males. They solicit her to follow them. And she complies.

But when they enter the thickest vegetation, Lotty just happens not to emerge from the other side. When it comes to sex, as it often does, chimpanzees have rules — and rule breakers. Rule breakers know the rules. And they know *who* rules. And Lotty has her own agenda — and a few tricks.

Ben calls from beyond the thickets. Lotty remains quiet. Everyone here is keeping

518

mum. No one is responding to Ben. No one is making a peep. Lotty listens carefully. For now, she doesn't want Ben and Talisker to know where she is, but apparently she wants to be close enough to them that they could hear her if she screamed for help. Chimpanzees know that others can be listening, and someone being hassled may exaggerate their screaming if they know that an ally will hear.

Because Lotty is now in peak estrus, Ben, as the top dominant male, wants to monopolize her. But Lotty doesn't feel like being monopolized — not yet. She is being cagey. Lotty has something going on with Daudi, even though he is only thirteen years old. This perplexes us as much as it likely vexes Ben.

"She's got a nervous scream bubbling up," Cat observes. "But you can see from the tension in her lips and face and voice that she's suppressing it." Lotty doesn't want to accidentally summon Ben. Chimps can give the impression that raw emotions drive them. Closer analysis, however, reveals near-continual filtering and self-control. In a social system that includes coercion and punishment, neither chimpanzees nor humans can afford allowing their emotions to rule without considering the consequences.

Lotty moves to Daudi and presents herself. Daudi inspects her swelling as though it's a crystal ball, holding, gazing, jiggling —. He does not copulate. You'd think he wouldn't look a gift horse in the mouth. But he risks getting beaten by Ben. He might be trying — in his young and inept way — to coax her away from the group.

Fat chance. She's not inclined to go anywhere with him.

Ben sweeps in. He's looking for Lotty. So, Lotty *chases Daudi*. She risks Ben's wrath if he thinks she's got something going on. So she makes it look like a public indictment of Daudi, as if saying, "How *dare* you assume liberties with me!"

Only now Ben is chasing Lotty, and not in a friendly way. Her screams trigger immediate screaming from the whole gallery. Then, as fast as it erupts, it quells.

Ben wants Lotty to follow as he leaves. But again she stays.

A few minutes later, Lotty "leaf drops" for Macallan, who is sitting right in front of me.

Cat is astonished. Leaf dropping, which bonobos occasionally do, has *never* before been reported for chimps. Unlike leaf clipping, which is noisy and obvious and meant to get another's attention, leaf dropping is silent and discreet, like a Victorian woman letting her handkerchief fall.

We remain perplexed at Lotty's choices. Lotty is a mature adult who has had kids. It's not clear why she's dallying with young Daudi and twenty-year-old Macallan. She could do better. Macallan can't. He obliges.

Considering Lotty's in-between behavior, I ask Cat Freud's question: "So — what do women want?"

"Choice," she instantly answers. "We want what we want without asking permission."

Why didn't Freud simply ask a woman?

"But," I say, "Lotty seems conflicted."

"That's the other possibility," Cat acknowledges. If Lotty goes with Talisker and Ben, she gets sex with proven males. But they will monopolize her. If she's here, she's among up-and-coming males, potential alphas of the future whose genes are also of good quality. "She has an interesting dilemma," Cat evaluates. "You could look at Lotty and say, 'First she was coerced into following two high-ranking males, then she was threatened by all these boys; clearly she's been manipulated and males control everything.' But I don't think that's the case. She might see it as: 'I dropped away from the two high-ranking males to have my choice among many.' She exercised her choice by staying back, disappearing from Talisker and Ben and staying quiet when they called." Her choices are harder to see scientifically because "you don't have a behavior to measure; she exercises her choice by *not* doing what *they* want."

Males exert power by force and intimidation. Females often exert their own power by what they choose to do or not do. "Perhaps this is what they're stuck with," Cat posits, "given the tendency of some males to hit first and think later."

A while later, Lotty walks away. Cat and I follow her into viny thickets, with some difficulty. I step on a swarm of army ants. Cat helps sweep off the dozens of aggressive ants instantly running over my pants. She tells me that this happened to her once while she was crawling through a thicket on her hands and knees. "Hopeless. I basically had to duck around a tree, get naked, and shake them all out of my clothes." Amusing. But this is not: this morning, I forgot to pull my socks over my pant cuffs — with consequences. With astonishing swiftness, ants are inside, up my legs, under my shirt, and on my scalp. And they're biting. I see one, on the outside of my sock, gouging, gouging. And this is what they're doing on my skin, inside my clothes. But so intent on biting are they that I can locate and squish them just by pressing a finger on my clothing until I feel a little crunch. That's what they get.

Cat suddenly thinks she's heard a "hunting bark."

Lotty, perhaps hearing the same, has *now* chosen to head to where the two big boys have been. As Lotty walks, the heretofore unloved Masariki, so ignominiously spurned by Monika, hurries ahead, turns to face her, stands upright, and opens his arms wide, offering her a full-erection invitation. Lotty just looks at him and walks past.

In this forest of green and shadow, crimson is not a common color. But Talisker and Ben have a young blue duiker, and they've torn the fawn right in half. Talisker holds the forequarters; Ben, the rear half. The sharing is political, a memorandum of understanding that they will support one another if either is challenged.

Lotty walks up to Ben and turns to present herself. Does she want sex — or does she want meat? She backs into him. There is nothing debonair about it; he complies with the fawn's hindquarters hanging from his mouth. If this harkens to the evolutionary origin of a human male picking up the check on a dinner date while the night is young, well, we've come a long way.

But Ben, who has not built a reputation on charm, doesn't share his meal. Maybe Lotty knows not to expect much from him. She does not bother begging.

Instead, Lotty begs from Talisker. She doesn't put her hand out, as chimps often do. She just sits a couple of feet away, staring at him. With an air of expectation. He tries ignoring her. She persists. He bites a chunk of meat and spits it out for her.

Meat sharing isn't generous or accidental. As with humans, chimpanzees usually share meat with: relatives, allies, potential sex partners. Rivals and competitors get snubbed. A male on the rise might seem liberal with his sharing. But when he reaches the alpha rank he's been pursuing, his seeming generosity shrinks to his political base, those who support keeping him in office. Humans in office hand out political favors. For chimpanzees, meat is often a political favor.

For human tribal hunter-gatherers (and even, to a certain extent, in modern societies), meat tends to be a different thing for males and females. For males, hunting enhances status, reinforces social bonds, and cements their family role as a provider. For females, meat is often a male-hunted, male-delivered source of nutrient-rich food, often crucial for the survival of her children, and it helps maintain the male-female relationship that involves sex. This dynamic has parallels in chimpanzees. Male humans like to brag about what they bag. Male chimpanzees often initiate hunting if their party includes a female with sexual swellings. In both species, meat is a prize.

Lotty picks up the meat. She begins alternating bites of meat with mouthfuls of a certain kind of leaf eaten only *with* meat. Chimps often eat leaves, but they're choosy, and they eat them at different times and for different reasons. Certain young leaves are simply food. Other leaves, though, are medicinal. Often first thing in the morning, on an empty stomach, Budongo chimps will pick a few rough leaves of *Aneilema aequinoctiale,* carefully position them in their mouth, then swallow them, whole. Intestinal worms get dragged along by the tiny Velcro-like hooks on the leaf surfaces, and pooped out. The plants chimps use for self-medicating differ from site to site, so this too is not instinctive; it's cultural. Young chimps learn which plants to use by — naturally enough — watching their mothers.

Lotty wants more meat. Talisker drops a piece but instantly moves it closer to himself. Seemingly that was just an accident; he has no intent to share further.

In her next actions, Lotty seems to show that she is particularly annoyed with Talisker. She looks to Ben to be sure he's watching the whole interaction. Then she gets up, gives Ben a bare-teeth smile of submission, and kisses and embraces him. She has ingratiated herself, with Ben's support, for — what? What does Lotty have in mind?

She goes back to Talisker and begs for more. He ignores her. She persists. He continues to ignore her. Lotty suddenly *chases* Talisker out of the immediate area — an outrageously nervy thing for a female to do. Ben doesn't intervene. Lotty circles back to hang out with Ben.

Before Lotty could press and perhaps aggress against the second-highest-ranking individual in the community, she first needed to neutralize the highest-ranker — Ben — so he'd sit it out. Lotty seems to have been playing a pretty shrewd game today.

Ben resumes ripping muscle out of the fawn's hindquarters. His lips, even his teeth, are crimson with the fawn's blood. He suddenly seems darkly transformed, a more fearsome being. Lotty looks directly into Ben's face while he's chewing. He moves off. Lotty and her six-year-old daughter, Liz — who has her mother's honey-colored eyes — follow close behind. Ben sits; so does Liz. Ben half-turns away from Liz and continues gnawing the fawn. Lotty positions herself out of sight, just behind a big fig tree, eating more leaves. When she hears Ben drop a piece of meat, she comes around. But Liz, who'd been sitting right next to Ben, has already picked it up.

Lotty tries a sneaky reach to take her daughter's meat. That doesn't work. So she tries grooming her daughter, in hopes of getting a share. She just happens to be holding on to Liz's foot, so her daughter can't go anywhere.

"Here comes Masariki," Cat says. "No meat and no girl."

Liz suddenly lurches out of her mother's grip — still in possession of her piece of meat.

By having dropped a little bit of meat for

Liz, who is just a pale-faced kid, and letting her and Lotty fight for it, Ben has succeeded in getting rid of both of them. He enjoys the remainder of his meal in peace. Rank has its privileges. Especially if you're smart about it.

The alpha *gets* meat, the alpha *gets* sex — and sometimes both at the same time. He displaces, and gains access. But what I can't see, and what I really want to understand is: Does the alpha *provide* value to the group? Does the system of male hierarchy bestow benefits to the community?

Most males suffer the indignity of lower rank. Few will ever become alpha. You'd expect the low-rankers to leave, and the system to have broken down a long time ago. How did such a seemingly unfair system start? Why does such a system persist? How does it self-perpetuate? So few get so much more than so many; what holds them all together?

All adult male chimpanzees engage in territory defense and physically risky aggression against adjacent chimp communities. The number of adult males — regardless of rank — determines how much territory a community holds. Sonso's territory shrank substantially when its number of adult males declined to a mere six. When maturing males increased the total number to about ten, Cat watched the Sonso territory expand. By keeping chimps from other communities out, community defense helps protect the community's food supply.

But even if male numbers are related to success in territorial defense, that still doesn't require a system of enforced male *rank* and *hierarchy*.

Cat sheds some light: "When there's a shake-up, when an alpha is being deposed — about every five to seven years — for a while there is no clear alpha. No one pant-grunts to anyone; everybody puts respect on ice until it's sorted."

In Sonso during one such turnover, skirmishes between evenly matched contestants simmered for four years, a relative eternity. "It was *very* disruptive to social life," recalls Cat. "A community without a clear alpha feels like it's not cohesive. They tend not to patrol their boundaries or to maintain other usual aspects of community life. So there's an order-keeping aspect to male dominance — even though in bonobos, again, females are the dominant ones who keep order, and they do it without the drama and violence that high testosterone makes males prone to."

Chimp alpha males can be bullies, but an alpha can also be, says Frans de Waal, "the consoler in chief; the peacemaker." They can protect underdogs, soothe and comfort distressed individuals, and create group coherence — more than anyone else. So there are two sides to alphas.

Ironically, then, the same obsession with status that causes so much strife creates the hierarchy that — most of the time — helps chimpanzees keep a lid on things. When you're living in a close-knit group, strife happens. That's not surprising, and so, in a way, that's not important.

What's important: that chimps have skills to deal with the inevitable strife. They come equipped with a sense of fairness, and they reconcile. Reconciliation, forgiveness — this is the path *back* from the brink. It's what holds the center, *creates* peace when peace is needed, and *maintains* peace when peace is in jeopardy.

Liz goes back to Lotty and signals a big scratch. She wants to smooth things over after their earlier tiff over the piece of meat. They groom briefly, then play a bit. Touch heals.

When Lotty moves off, Ben and Talisker shadow her closely. They are getting more and more fixated on her — and on each other. All day, whenever Talisker has sat, Ben has made sure to position himself at least as close to Lotty, or a little closer. It doesn't look like much is going on, but they're subtly testing and reasserting their boundary. And with frequent bouts of mutual grooming, Talisker and Ben are also continually bringing their rolling boil back down to a simmer, reducing their tensions, in one of the most competitive situations two male chimpanzees can find themselves stuck in.

During our evolution, too, gaining a mating advantage drove the effort and risks of competing for male status and rank. Workplace harassment patterns indicate that not a lot has changed; human males too often feel that high rank entitles sex. And sometimes, as with alpha chimps who are bullies, like Nick was, the higher they rise, the harder they fall.

Macallan and Daudi come to greet Ben. "They did that because that's what you have to do," Cat says. "They're interested only in Lotty."

The two sit near Lotty. Macallan gives a long stretch of his arm and ever so slightly rolls his body; he's getting closer to Lotty while trying to mask his plan. Ben, who's no dummy, moves in Macallan's direction. That's all it takes.

Everyone gets up. Their procession resumes. On most days at this time, they'd probably visit some fig trees that are ten minutes away. But Cat says, "Here's a surer bet: whatever Lotty wants to do, the rest will want to do. Today the world revolves around her."

Lotty, suppressing a tight whimper, keeps dropping to the back of the group.

Lotty stops.

Everybody stops.

Lotty climbs. Talisker climbs, of course. Ben

chases everybody else away, then climbs above Talisker.

Ben extends his right arm to a leafy branch and gives it a vigorous shake. Alpha males give and receive more gestures than other chimps, but Lotty, sitting a few branches above, does not respond to Ben's invitation.

Lotty descends. Ben and Talisker descend. Ben leaf-clips before heading into a thicket. This time, apparently in a changed mood, Lotty follows.

Finally, the stars and an ovum having aligned, Ben copulates with a receptive, fertile Lotty. He wipes his penis with a wad of leaves, which he sniffs.

PEACE:
NINE

On this new day, Ben, the alpha, and Lotty, the senior female, seem absorbed in the mutual pleasures of grooming. But with every distant call, both pause. There's never total relaxation because someone, somewhere, is doing — something. Monitoring chimp sounds coming from different directions in the forest is their version of scrolling through their phones, obsessively tracking who's doing what, where, with whom.

A couple of muffled hoots through the deep forest seem no reason to interrupt this peaceful private groom. But no matter the placid moment, you never know when the next altercation will erupt (though you can be sure it will). One call in particular snaps everyone to attention. Lotty leaps up and yells, *"Wah!"* She hurries forward a few paces. Ben's right behind her. He's puffed up in case there's trouble. But he doesn't want Lotty to leave. He gives her an open-mouthed kiss on the back, then embraces her, then resumes grooming her.

Underneath the returning calm, something's brewing. Far-off chimp calls arrive dissipated and soft through the filtering forest. But Lotty again immediately responds with short, sharp screams — loud — that draw out into a sustained wail. *"Wah! A-weeh. A-waaagggh!"*

Cat's perplexed. "It doesn't make any sense for her to respond so strongly and with such apparent fear to such distant calls," she says, "unless she hears that something is happening to someone she's close to — her daughter Liz perhaps — or she's been agitated by something we missed early today, perhaps an earlier encounter with the other community."

We hear chimps to our left. They're answered by chimps to our right. Not close. Cat thinks those on our left are from the adjacent community, an unstudied, slightly mysterious group referred to simply as the Easterners. But at this distance, she can't entirely distinguish their calls. Different individuals *and* different communities have different-sounding pant-hoots. The Waibirans' pant-hoots have a big buildup and a long letdown; in the Sonso community, the beginning and ending of the call is more abrupt. A few days ago, one of the grad students asked, "Do the Gombe chimps sound really different from ours, or does Jane Goodall do a really bad pant-hoot?" Answer:

they sound really different. Especially to chimpanzees.

Different dialects, different accents, different voices. Because: different individuals, different learned customs. Chimpanzees vary, and chimpanzee culture is variable at every level.

Kizza concludes that those to our left are indeed chimpanzees of the adjacent community — the enemy — calling from the boundary area. Chimp territorial boundaries are as real as in human tribal societies.

Cat assesses the sounds: "It's a big group over there." The party we're with is small.

Most intercommunity interactions are hostile shout-offs that can last minutes, even hours.

But sometimes, it's war. And near borders, the stakes are high. Don't be caught without major backup. Attackers tear off ears and bite off fingers and testicles, the kinds of mutilations that, in humans, happen only in the worst gang wars and hate crimes. Females, too, can suffer severe beatings. No one is immune.

To make war, you must have a strong sense of belonging to your group. You must subjugate your sense of individual self and personal safety. You must erase the individuality of the enemy. Humans dehumanize other humans as one prerequisite for warfare. In the mid-1970s, during her classic research at Gombe,

Jane Goodall documented a long, unfolding event in which some of the chimps in the Gombe community split off and moved south, becoming a new community. In what became known as the Four-Year War, males of the original group banded together to hunt down the separatist males one by one, savagely killing them.

About a tenth of adult male chimpanzees die in warfare. Surprisingly, mortality rates due to warfare in many *human* societies have often been twice that. Most human societies exist in some continual state of war. In New Guinea, warfare in many tribes accounted for 25 to 30 percent of male deaths; but in parts of Europe — for example, Montenegro in the early half of the 1900s — deaths from war reached about 25 percent of *all* adults.

Lotty had screamed; it might have been better if Lotty had stayed quiet.

"It's not clear to me," Cat offers, "whether Ben wanted just to reassure her or wanted to quiet her screams." Many times, Cat has seen a chimpanzee scream when silence would have been smarter. For instance, a female might want to steal some meat from a male who is sleeping. She's right there, all she has to do is reach out, but she lets a little sound of excitement bubble out of her — almost reflexively — blowing her own cover. Suppressing the expression of fear seems really difficult for chimpanzees — even when it gets them into more trouble than staying quiet would.

The Waibira chimps' thirty adult males represent such an intimidating force that there's no downside to letting enemies know that their highest-ranking chimps are within earshot. But all week they've responded to rivals' challenges mostly with silence. With so many Waibirans sick from the coughing cold during the last couple of weeks, Ben and this group have been hesitant to engage in territorial challenges that they'd normally respond to quickly. A lot of energy goes into responding. It's highly stressful to be in contact with another community that might be encroaching into your territory.

"I think many of the males haven't been well enough to take on the challengers," Cat reiterates. "They just weren't up to a confrontation." And so: Eastern chimps have pushed into Waibira territory.

There's now less coughing, and it's drier-sounding. This is the first day that Ben seems motivated to respond affirmatively. He also appears conflicted; he's still not entirely well.

"If I were Ben and Lotty," Cat speculates, "I wouldn't want to act without a lot more of the big boys around me."

Ben and Lotty and the others get up. Lotty disappears.

Whenever there is a loud call, the whole

forest lights up with ringing responses. Many others, whom we cannot see, are nearby.

Ben hesitates, then pivots, does a lot of to-ing and fro-ing.

Suddenly someone is walking downslope, silently. Ben looks carefully to get a visual ID. It's Lotty.

A little group assembles. With lots of pant-grunts and arm reaches, they're checking in with each other. Cat says they're "psyching themselves up a little bit to keep the group tight as a unit."

Gerald arrives. And Monika. When they all seem to feel they have a sufficient critical mass of males, they simply get up — and go.

They lead us up a steep, rocky pitch to a hill beneath gigantic trees. The undergrowth opens. I wonder aloud what's on their minds.

"If I was a chimp," offers Cat, "I'd be thinking, 'If I'm going to recruit anyone else, I need to do it now. Over this hill, no one behind will hear me.' "

Ben, hair erect and agitated, loudly roars, *"Ahhh. Agghhh. Agghhh. Agghhh."* It's lower than any sound I've yet heard him make, more aggressive, threatening. Actually a bit frightening. He begins charging around, hooting and thumping his feet hard on the boomy buttress roots. He listens. Repeats. He's trying to raise a bigger militia.

Chimps who hear his buttress banging will know it's him. Even *that* is individualized.

Cat and our guides know whose drumming they are hearing by sound alone. Some use feet only. Some add hands. "One chimp does two feet plus one hand. Another is free jazz," Cat says.

Sam arrives, and one-legged Tatu on the crutches of her long, strong arms. Alf and Lafroig appear. Their silence speaks of tension, apprehension. Several climb into trees and now — "Woough!" All join in: "Ooh waagh." "Ouwww." "Whooohh." "Hoo-ahhh, hooohh ahh —." "Ahhhh wahhhhh."

For Waibira members, this means "Come here; we are rallying." At the same time, it says to the other community, "We are many. We are standing our ground."

Because so many Waibirans are still a bit sick, it might be a bit of a bluff. Perhaps they intend that.

With points made, everyone subsides to edgy silence.

Sudden distant drum-thumps and hooting ringing through the forest instantly make all the males puff up.

"They seem constantly at a low-level state of war," I say. "Long moments of relaxation punctuated by —"

"They're never fully relaxed," Cat counters. "They're always just a bit on edge."

The adjacent Easterner community calls from near their shared border. Waibira responds — loudly. Essentially all the adult males, and any females without young, begin to hurl their vocal strength at their distant rivals. Choruses rise on either side of the border.

The volume reflects the number of callers. Numbers determine whether there will be physical contact, and whether the fighting will turn lethal.

Another round of distant calls sets off an absolutely fever-pitched eruption of hooting and shrieking from our group. Throughout all leafy points of the compass, the air reverberates. Those above scurry out of the trees like firefighters who'd been dozing one moment and in the next are summoned into action, sliding down toward the emergency.

Suddenly our chimps are gliding toward the calling Easterners across the sun-dappled forest floor. But they are *q u i e t.* Audio blackout.

Moving absolutely silently, walking with almost no noise in dry leaves, a long caravan of chimps travels for ten or fifteen minutes on one of the major trails. With palpable tension, they seem ready for combat. It *feels* like war.

One of the chimps is so nervous about the prospects for an upcoming encounter that he stops and has diarrhea.

I'm following behind Gerald. Cat explains, "One of the big males will go first; another might go last. Another may act almost as a guard, hurrying them all."

In contrast to their silence, our footsteps are so loud in the dry leaves that we are spoiling whatever element of surprise they might have wanted.

No, Cat says; I have that wrong. Notice, she says: they keep stopping to wait for us. "They're good tool users, and they're using us. The other community is afraid of humans. Just the sound of us will halt their advance. They might even retreat."

On the prolonged march toward possible contact, one chimp picks up a small stick and pauses to sniff it for clues. All stop, and each sniffs one particular branch. The rivals were just here. Now our group moves off at a changed pace, very much slower and more deliberate; they're apprehensive, hesitant, hunting for those who may be hunting them.

Masariki flashes a fear grimace to Gerald. They briefly hug mutual reassurance. Then swiftly hurry to rejoin the others.

Cat says of poor Masariki, "He's probably pretty terrified." They seem frightened by the *thought* of the violence that awaits. There's a lot to worry about. "When you've got three or four big chimps holding one guy down on the ground, ripping off his testicles, breaking and biting his fingers, biting around his throat —." They'll beat their captive until he's unresponsive. "Then they'll sit," Cat continues, "and watch, very closely. Any sign of movement, any sign of breathing, and they're right back at him until they're satisfied that he is dead. So, if you've wondered whether they understand death —"

For about twenty minutes, the chimps had been covering ground in defense of their homeland. Now they've pushed beyond the boundary, incurring about a third of a mile into Easterner territory.

Everyone halts. Exquisitely attentive. One chimp stands upright on his legs, putting one hand against a tree trunk, peering ahead. There are brief touches for reassurance. Then all advance, with silence so absolute it feels like everyone is holding their breath.

Shouting from the challenging community accelerates our chimps. Still in silence, the Waibirans burst into a running charge.

Contact!

A great racket breaks out, with terrifyingly guttural roars and thunderous stomping. Ben and Talisker come barreling along in full display. Other chimps show strength by breaking off the biggest branches they can manage, dragging them around, and throwing things in their enemies' direction. Screaming Easterners start streaking away in retreat like dark comets, leaving the dense vegetation shaking.

Waibira members are yelling a wide array

of hoots, wahs, and screams — long, confident, and assertive. Distant alarm calls ringing through the forest inform all that the other community has drawn back.

Suddenly we realize that four or five female Easterners are silently stranded in the trees overhead. Our chimps seem intent on trapping them. This is no game. Any female caught will get attacked, possibly fatally.

We have an ethical problem now. The Eastern strangers will not come to aid their stranded members while we are here. Not only are we giving Waibira chimps a boundary-line advantage; we are giving those trapped females a dangerous disadvantage. They cannot recruit help. If they get beaten or killed, we, by our mere presence, will bear some responsibility.

But those stranded females are a large enough group to dissuade the Waibira chimps from risking all-out fighting high in trees. So the Easterners begin working out their escape through the canopy, moving away, tree to tree.

However — one of their females, marooned alone in a tree near the skirmish line, decides to bolt to the ground and make a run for it. Gerald rushes her and seems to manage a bite but fails to grab and tackle her. She escapes.

Peace:
Ten

A substantial group of Waibira chimps seems to have us encircled. We suddenly realize there are about twenty; some are just out of sight in dense vegetation. We find ourselves surrounded. But there's no problem at all; we've simply walked right into a group that's resting. They remain relaxed as we join them in sitting on the ground. We, too, can use a rest.

It seems a fitting time for a snack, so I pull out my bag of raisins and almonds. Kizza is sitting next to me, and when I offer him some almonds, he looks at them suspiciously.

"What is it?" he asks.

"They're called almonds. They are nuts."

"You brought with you?" he says as he tentatively accepts my offering. "Or you bought in Uganda?"

"I bought them at Lucky Seven, right in Masindi."

"I have never seen. How much?"

I'm embarrassed that I don't recall. I was

simply in the mood for the bag of almonds and a package of dried dates. I didn't look at the price. (Thirty-three hundred Ugandan shillings equals one U.S. dollar.) Now I realize that for Kizza, a bag of almonds would be out of orbit. Kizza is thirty-three years old and has been working here on the Budongo chimpanzee project for five years. Before that he worked for the National Forestry Authority to help protect the forest itself. He's got a bachelor's degree but could go no higher, he says, "because, no money." When not in the forest with the chimps, he has side businesses at home: "Shaving heads. Also digging." He explains to me, "In Uganda you must dig, to grow maize, potatoes; to get money for salt, and for the school fees." His schoolgirls are twin twelve-year-old daughters and a third, age eight. We are sitting shoulder to shoulder, but the distance between our cultural circumstances is enormous.

Kizza, who has never been beyond Uganda, asks me, "Is Uganda beautiful?" He seeks my perspective as an outsider who has visited other lands.

I tell him yes, Uganda is beautiful. And he and I are both satisfied with that answer.

For a few moments, we sit in silence. For another hour, the chimpanzees relax, many sprawled on the ground in repose. For this while, there's sufficient peace for them to rest and quietly socialize. I listen to the forest

544

birds until, in the tropical heat of midday, the coos and whistles and trills quiet as heat brings on an abundance of green silence. There is neither buzz of insect nor flute of bird. Even the butterflies still their flutters in the peak of the heat. In a quiet that is absolute, we sit eased into stillness.

Just fifteen feet above my head, in the daybed she has just constructed, the female named Bahati lies with her two-year-old son, Brian. Nap time.

Kizza rests his head on his pack.

My mind slows to the pace of the parallel lives around me. All of us are slipping through the same hourglass, each somehow on a different trajectory, unfathomably allotted a disparate span. Sitting in the leaves with my back against a tree, I let my eyes flicker closed.

For some minor eternity, I sleep. Steeped in stillness, I absorb the timeless otherworldliness of this place, so far beyond my accustomed civil circumstances, out of reach of everything familiar.

When I come to wakefulness, it's quite as if, under a magic spell of time travel, I had dozed off in the twenty-first century and awakened five million years ago in an unbroken infinity of forest in a prehuman world with a canopy of dark apes against the sky.

I turn my head to see Cat sitting twenty feet away, scrolling through her cell phone.

And the magic spell, like so much else, comes undone and breaks.

All reverie definitively ends when calls from afar trigger a round of hoots, grunts, waah barks, and screaming. Lotty listens intensely, assessing her familiars and any strangers. This is the flow and ebb of chimpanzee days, the tidal risings and subsidence of chimp emotions. Collectively their stories braid, their stories fan apart. They pause and resume. They have no beginning, middle, or end — for now. Life goes on in a circular way. The time of day, the arc of individuals, their years and their careers — the seasons of their lives as they rise and fall and vanish into the dust upon which each new generation travels.

"Uh-oh," Cat says, a bit alarmingly. "It's going to be fluffy chaos."

Chaos?

"We've got all the little ones coming."

Forty-year-old Kidepo arrives, her infant's tiny hands gripping her belly. Ndito-Eve comes with seven-year-old Noah and bouncy two-year-old Nimba. Most chimps have dark eyes, but Ndito-Eve's left eye has a white sclera like a human's, making her gaze quite striking. Bahati strides down with little Brian riding her back. Her little cowboy hops off momentarily when she pauses on the ground to assess the scene. When he taps her to indicate that he'd like to get back up, she dips her shoulder, telling him to hop on.

Brian jumps from her back to a low branch, climbs a bit, and repeatedly dives into the springy canopy of a small tree. He and a couple of others start chasing each other up and down vine-tangled trunks, then jump onto Monika, who eagerly rolls around with them, tickling, pulling their feet gently, and making open-mouthed contact.

An open mouth with lips covering the teeth is called a play face. "The lips start to get pulled back as they start to laugh," Cat explains. The kids are chuckling, their laughter sounding like rapid panting. Human laughing originates from the rapid play-panting evident in our ape cousins. We think of laughing and smiling as the same kind of thing, and the smile-like play face escalates into laughing. But laughing's origin in play-panting differs from social smiling's roots in the bare-teeth grin that signals nonaggression.

Play strengthens the bonds that help define and maintain social groups. Phyllis Lee, who has studied elephants for decades, tells me that individuals who are more playful are more likely to survive. So play is serious stuff. But it must not be done seriously. Each participant must signal that this is only play. It must feel good, not frightening. And so to go along with this serious practice that gets done non-seriously, evolution provides an informing emotion: a sense of fun.

The ability to have fun is an evolved capacity; its origins precede humans. The neuroscientist Jaak Panksepp was greatly ridiculed for claiming that his lab rats loved being tickled and that they laughed at frequencies too high for the human ear. But by recording his rats laughing and then downshifting the frequencies to within human hearing and thereby providing proof, Panksepp got the last laugh. My wife, Patricia, and I raised an orphaned squirrel who loved being tickled and would roll over so we could get to her belly while she squirmed, kicked, and nibbled our fingers. She'd come back for more and more until we had to break off the play session and get on with our lives. Young raccoons, too, love being tickled, though as they grow their idea of play can get too rough for human fun. Owls are predatory birds, but people lucky enough to raise an orphan or work at a nature center may realize that owls enjoy getting their heads scratched and their beaks rubbed; young owls like to pounce and play with toys, similar to cats and puppies. Many creatures play, and what motivates them in the moment is that they have fun doing so.

But as with human children, chimpanzees sometimes get carried away. It can get rough. When one of the youngsters starts screaming in fear, their mother rushes in, also screaming, to break up the roughhousing. She reprimands the unruly kid. But that kid's mother objects, and a fight erupts between the mothers. For a few seconds, the forest sounds as if the gates of hell have burst open. Chimpanzees fleeing the fighting are streaking off in all directions.

Just when the screaming and hooting escalate to a fever intensity, Ursus, a bear of a chimp three decades old, rushes in. Wielding his relative authority with a major display, he seems determined to break up the riot and quell this round of trouble. It works. Ursus has brought calm as if by decree. Order returns. Social difficulties are inevitable; restoring peace takes effort and skill. Male chimps can both disturb and restore the peace, be peace breakers and peace brokers. For that to work, a chimp like Ursus has to understand this, want this, and know how to make it happen. That seems to be exactly what we've just watched.

Ursus is the biggest and strongest guy here at the moment. But he's mellow, not competitive; he doesn't seem to be gunning for alpha. "Maybe he'll do the strategy of 'I'll just hang here with the ladies, father a few children, live a fat and happy long life,' " Cat speculates. That's the strategy used by a chimp named Pascal over in the Sonso community. Pascal, says Cat, is a "total ladies' man." Talisker, too, as we've seen, can sit quietly off to the side, not displaying to attract attention, yet many females go out of their way to greet this elder statesman.

If Talisker was once the alpha, as Cat has said she suspects (and even if he wasn't), it's often obvious that he's carved out for himself his unusual role outside the competition and the hierarchy, that of, simply, a well-respected elder. One of the ingredients of his success, Cat observes, is that "Talisker has maintained a diverse social network." Whether that's due to strategy or just longevity, we don't know. Ursus is also interesting. He seems to be a point of light in this community, a way of being male without the drama, deriving respect through seniority. One might say he's earned it rather than won it. Even in chimps, a peaceful path is possible. And even some male chimps travel it. And when they do, sometimes they come into a fight and use their authority to get everyone's attention and make peace happen.

Meanwhile, four-year-old Nalala is spinning one-armed on a vine. His mother, Nora, raises an arm to say "Come," and pulls him into her embrace. Their contortions are comic, relaxed. She rolls him onto his back, buries her face in his belly, and plants a series of open-mouthed kisses on it, like a human mother blowing raspberries on the belly of a giggling infant.

Nalala means "sleepy." But now he is hungry. It's the dry season, the weaning season. Nalala makes frequent requests to nurse, using strings of gestures directed toward his mom. He taps his mother with his foot. Tap tap tap —. He's persisting, and he's elaborating. Back and forth between them, it's a little negotiation. With her own fingers, Nora taps him on the back, meaning, "Okay, c'mon — let's try." Nora could just pull him to her, but instead she communicates about it. With his request obliged, Nalala tries one nipple and then the other, then backs off.

Nora is depleted. Nalala begs again. He uses arm raising to indicate, "I want to nurse." An arm raise generally means "Come closer." Context informs the finer meaning; they both know why he's requesting the intimate proximity.

Right now, Nora's face bears the expression of a mother with an irritated baby. Nalala suddenly erupts into an absolutely pitched weaning tantrum, facing his mother while shrieking, hitting, and even slapping her in the face.

Nora neither retaliates nor complies. She merely hugs her child closer, which tempers his tantrum. Eventually they relax, lying along a thick vine not far off the ground, with him on top of her. Mom wants to nap. He keeps whimpering. Every now and then she opens an eye, then goes back to trying — perhaps pretending — to nap. He pitches a smaller fit, interrupting himself partway through it to see whether his mother is paying any attention, and whether it's worth his effort to continue the performance.

Nora is a confident, socially versatile mother. She's not nervous when she's alone. She also often hangs out with big, high-ranking males. That means that from birth, Nalala has been building bonds and watching their dynamics. "And *that* means," Cat tells me as I watch him not-sleeping with his mom, that "Nalala will be an outstanding contender to become a very high-ranking male."

He's such an innocent and delightful little bundle. It's unpleasant to think of him three decades from now, locked into a lifetime of competitions, fights, and threatening displays of strength. Being female comes with significant drawbacks. But so does being male in a male-dominated world, because to be male is both to be dominated and to incur the costs of dominating. Maybe he'll choose the path of Ursus and Pascal.

It's hard to predict what his future holds. Chimpanzee lives run a colorful palette. We've seen that some males prefer a peaceful life and stay out of the political scrambles, while others are obsessed with status and strategic alliances. Females are less political, and females fight less. They form friendships, but not to gain political advantage; it's just generally better to travel with a friend than alone.

Conflicts within social groups are inevitable, so managing them is crucial for group stability. High-ranking chimpanzees of either sex sometimes intervene to quell fights. Such impartial intervention by bystanders reflects a sense of community concern that is rare among non-humans (and not universal in humans). Acting as if fights are considered "bad" suggests a moral nature, a *sense* of good and bad.

Because conflicts do inevitably arise, reconciliation *after* fights is very important. Males are both more aggressive *and* more conciliatory. Males have more to gain and lose from one another's support. They must cooperate in hunting and territorial defense. Sometimes a third chimpanzee will get between two chimps who need to patch up a fight, and will mediate. The intervener will start to groom one of them, so that soon they are both grooming the mediator. Such a mediator understands the relationship between the two others, a cognitive level called "triadic awareness." The mediator might then get up and walk away, leaving the combatants in a situation already cooled down. Then the combatants will groom each other, smooth things over, and be all right. None of this is accidental. Because it takes concerted effort and a plan, it can happen only if this is their *desired* outcome.

Nambi — at age fifty-six a very old lady for a free-living chimpanzee — used triadic awareness to engineer her son Musa's rise in rank. Nambi would groom dominant males, then groom her son. "So everybody would shuffle together and kind of be squished up," says Cat. Nambi would then move out of the cluster, leaving her son interacting in a high-level grooming clique.

Chimpanzees hurt and help each other, compete with and assist one another, because they are intelligent and aware of different layers of interests, of overlapping and even contradictory goals. They have the wits and the long-term memory to know who their friends are, with whom they have mated, and who is their bitter rival.

Chimps seem to understand that "love your enemy" is highly practical advice. Such effort is not simply a way of restoring temporary calm. It is a way to set things whole, a way to hit Reset on group identity and community belonging. So, too, with us. Revenge isn't the only way to settle scores, and it's not the fix that helps everyone. When there's a leak in the social flow, there is a need to right the boat, to keep social things social. Coins of empathy have different sides, and healing the harms we've caused is one of them. Deciding upon forgiveness and offering reconciliation, even aiding those with whom we've had sharp differences — such pro-social acts let us turn the page, move on, and — most crucially — keep things together. The lines between chimps and humans here are blurry, because the impulses are kindred and the remedies comparable.

Just as Margaret Mead identified "reciprocity" as a cornerstone of human society, Frans de Waal has identified "reconciliation" as a cornerstone of life in chimpanzee society. Humans did not invent reconciliation and peacemaking. We see a strong tendency to forgive and move on in other species. For many of us, it's visible most often in our dogs. Whenever their relationships hit a literal snarl, they don't want to hold a grudge, either between them or with us; they want to lick and make up because they know implicitly, deep down in their core being, that their relationships are everything and their reliance on them is total. The bear expert Ben Kilham has seen reconciliations in dozens of orphaned bears he has raised and rewilded, where squabbles are frequent but the need to maintain a social network is crucial. Reconciliation probably exists in all of the mammals for whom dustups and disagreements must be fixed so that life as a group or community can go on. And it appears to be there in the parrots, with their mammal-like capacity for rage and their propensity for mutual preening and tenderness.

Following a fight in a zoo's chimpanzee night quarters, the attacker spent a large part of the day tending his victim's wounds. Such behavior seems to verge on a sense of responsibility, perhaps remorse.

Is guilt, shame, or remorse possible in nonhuman beings? "What feeds guilt and shame is a desire to belong," writes de Waal. "The greatest underlying worry is rejection by the group."

Group identity — such a fundamental aspect of culture — both enables and results from empathy, altruism, cooperation, and the need to keep things okay. The driving force is the need to make the group continue to cohere as a whole, because the benefits of the group are greater than the summed collection of its individuals.

Nora moves a couple of paces, stops, raises one foot, and wiggles it to get her baby to hop up. When Nalala gets on, he gives a little "Let's go" pull. As Nora and Nalala move off, Nora lets him belly-ride.

Cat says, "There's no way a chimp that age should be riding his mother's belly." When being weaned, young chimpanzees often revert to infantile behaviors. Though they've been walking and climbing on their own, like this guy, they might demand to ride like an infant. Human children also do some regressing when the growing gets tough.

In a rare patch of sunlight striking the forest floor, Nora stops. Nalala disconnects from her belly and climbs onto her shoulders, and they continue. He rides her into the forest, sitting upright like a mini-maharaja aboard an elephant, with one leg tucked under himself, seeming as comfortable and secure as any kid might, as if he's inherited the world.

PEACE:
ELEVEN

It's a few weeks into my Budongo sojourn, and today we're taking a break from the Waibira chimps — and vice versa. Our idea of a "break" is simply to visit the adjacent chimp community known as Sonso. Sonso's population is about half the size of Waibira's. And as we mentioned earlier, compared to Waibira's roughly even female-to-male ratio, Sonso is more typical, with roughly two females to every male.

Hawa is Sonso's twenty-five-year-old alpha. Hawa's mother, Harriet, is here right now, and I am surprised to see that she has a new seven-month-old baby. Harriet greets another veteran, Kigeri. Both are about forty years old.

A mother with a young baby will often go out of her way to make sure that everyone is good with her and that she is good with everyone. Like a political candidate, she is in a sense campaigning for her infant's safety, doing a little extra socializing to ensure good, harmonic amity all around.

Harriet makes a long reach to Kigeri, but rather than engage in a handshake, Kigeri puts her mouth on Harriet's hand. The reach says, "I will trust you not to bite my hand." Kigeri demonstrates that the trust is well placed: "See, you can trust me."

On a massive fallen tree, Oakland is playing with her bright-faced three-year-old, Ozzie. There's gentle open-mouthed play-biting, lots of reassuring arms around the baby.

Young chimpanzees play frenetically. But having them around has a calming effect on adults. Youngsters are like little bonding centers, often bringing adults together to join the play, creating trust, helping maintain community.

The great Musa arrives, but with no display or fuss. Melissa is right here with her two-year-old son, Muhumuza. Muhumuza attacks and play-bites Musa. Minutes of rolling, wrestling, and happy grunting ensue. When the baby finally sits up, the lordly Musa pokes him, just to resume the play.

Another adult — this one with a pale face — approaches and lies on her back near the baby, extends a hand, and, when that doesn't work, gives the baby a light tap to tease him into pouncing. Little Muhumuza pounces. The pale-faced provocateur covers her eyes to protect her face while flailing back with one arm at the attacking ball of fluff.

Simon, twenty-five, comes over from where he's been resting and gently grabs and lets go of the baby's feet. It's a tender moment.

Cat says, "These are the times when you forgive them for so often being bastards to each other."

In a surprise move, Muhumuza takes hold of Simon's penis and *hangs from it.*

"Not the smartest thing to do to a large male," Cat observes, "but little males do get fascinated by adults' penises." Simon lifts Muhumuza onto a branch and unwraps his hand as if to say, "No, that's not for you."

Babies get a free pass from many social norms and transgressions. One day the former alpha male Nick was copulating with one of the popular females when a youngster named Klaus ran up and just pushed him off. That was the last thing the most dominant male expected, but he did not attack. Klaus's mother, playing it safe nonetheless, came screaming in and grabbed her baby and kept running.

Males often appear to blow up for no reason. And then you see a male with a kid dangling from his penis, and he's very gentle and tolerant. Cat says, "Their combination of aggression and restraint seems an odd business. They're not all violence on a hair trigger; they *can* control their responses."

Yes, chimps can be violent, impressively so, and the violence can stay with you. "But, y'know," Cat points out, "if a young one has just screamed because a big male has frightened them, the male will sometimes run over and give them a kiss or a hug. That ability to instantly pivot from being huge and scary to incredibly gentle — it's equally impressive."

After the chimps have indulged in a long period of play, the little ones fidget while their parents try to nap.

It's all so peaceful.

Chimpanzees, writes Budongo's pioneering researcher, Vernon Reynolds, "have evolved a high degree of what can be called social intelligence, involving appeasement, deception, counter-deception, alliance formation, reconciliation after conflicts, and sympathetic consolation for victims of aggression." Theirs, he says, is a "thinking society," based on the intentions, plans, and strategies of its members. These are capacities we share, enabled by brain circuits we share, originating in evolutionary history we share. Chimpanzees provide both a cautionary tale for how to get it wrong and instructive mentoring for how to bring it back and make it right.

Being with chimpanzees creates a swirl of comparisons and assessments. They come up good and bad. So do we. When we measure them by our metrics, they never measure up. They are not us. If we measured ourselves by their calibrations, we'd see that we are the apes who have most intensified toolmaking, warfare, pursuit of status, oppression; that we are as trapped into enforcing and patrolling boundaries as they are. We would also see that we are by far more creative, compassionate, and communicative, as well as much kinder. Humans are both the most peaceful and compassionate *and* the most lethal and destructive species. Human anger and violence is not an aberration of a few marginal individuals; it's one characteristic component of our usual repertoire.

It is as hard for us to see chimpanzees for who they are as it is to see ourselves for who we are. We see chimpanzees by our own light. But by our own darkness we miss much. If we really understood who *we* are and how we might be, we'd realize that we have the choice to honor the compassion that is best in us and grow past what is most unfortunate. But we'd have to take a really good look in the mirror and decide — as only *we* might be capable of doing — what kind of beings we want to be.

In the weeks I've been here, we've often witnessed bluster and aggression. Like human societies, chimp societies have the potential for episodic extremes of violence that rivet us with fascination and horror. Such episodes in Waibira skewed my beginner's impression. The recent aggression level there is outside the norm, Cat says. She hopes

562

things will soon settle down.

Moreover — there is no norm. Says Cat, "The present uptick in violence doesn't represent 'chimpanzee behavior,' whatever that means." Individual personalities, changes in their environment, gender ratios, population pressures — as in humans, any of these might contribute to the tone of a community. "Chimps differ," Cat emphasizes, "both individually and group to group." For chimps, "difference" is the norm.

I've been trying to answer the question "What are chimpanzees like?" But as Cat has been trying to get me to see, chimps are not "like" one thing, in any one way. I've been looking *at* them. Cat sees *into* them. The question I needed first was "Who are chimpanzees?"

Cat has been trying all along to show me that the violence, the male ambition that has impressed and distressed me, is just one aspect of chimpanzee life. She's been trying to get me to understand that chimpanzee life is many things in many places, whether in different communities here or across Africa, and that the tenor and temper of chimpanzee communities, as in human communities, varies over distances and changes through time.

Over hundreds of thousands of years and into recent times, the four genetically differing kinds of chimpanzees have devised differing answers to the question of how to live where they live. Eastern chimpanzees, including the chimps here in Budongo, live in heavy forest. Western chimps live in a woodland-savanna mosaic; they sometimes use caves for resting and sleeping, craft those long daggers for hunting bush babies, eat far more termites than other chimps, play in water (though chimps cannot swim and most chimps are terrified of immersion), and will even travel and eat at night. Socially, Western chimps are more like bonobos. There aren't peripheral females; everyone hangs out together. They are more egalitarian. They more easily share meat without rank or politicking dominating who gets a share. And with only a couple of killings recorded in decades of study, their murder rate is so low it approaches bonobos' zero lethality. Chimpanzees don't "have culture." They have many cultures.

"Anyways," Cat adds, "it's still chimps' good side that I mainly see — even if in some years I need rose-tinted glasses to maintain my love of them."

So okay, I get it; there is no such creature as "the" chimpanzee. They don't live in just one way, or stay one way. Human groups differ in their cultures, and chimpanzees, as we've been seeing, also differ in theirs. What humans call "chimpanzees" are many creatures, living in many places, in many ways, over changing times. Their differing lives matter to them. They *should* matter to us.

The current time — the time of us — presents chimpanzees with their greatest challenge. Africa keeps the deepest of primate pasts. The question is whether it holds a future. The continent makes no promises. Neither do we. Chimpanzee pant-hoots no longer pierce the air in the nations of Benin, Togo, Burkina Faso, and Gambia. In Ghana, chimpanzee existence dangles by a thread. Forests continue to be lost to logging and farms, people kill chimps to eat them; there's the sordid primate pet trade, and the wide wreckings of human civil wars —

"Current threats and rates of declines," Cat notes, "are terrifying."

From the 1990s through the 20-naughties the number of chimpanzees in Côte d'Ivoire declined 90 percent. In the last forty years, chimpanzee and gorilla numbers have, overall, dropped by more than half. In the first two decades of this century, Borneo lost half its orangutan population; some one hundred thousand orangutans have died because agricultural corporations have destroyed their forest homes, largely for oil palm plantations. Needing fifteen years or so to become fertile, then giving birth once every four to six years, apes would need about 150 years for their populations to recover if all their problems stopped today and their habitats reappeared. Unlike the apes themselves, the problems aren't going away soon.

Jane Goodall's mentor, the great anthropologist Louis Leakey, famously observed, "We are the only animal that makes choices that are bad for our species." That's quite an understatement. There is nary *any* species our choices benefit. Albert Einstein overstated it when he said that humans are "endowed with just enough intelligence to be able to see clearly how utterly inadequate that intelligence is." Few see. The world that made us possible, we are making impossible. Among the withering lives at stake are the different kinds of chimpanzees, inventors and practitioners of different ways of living. Recently, scientists looked across all chimpanzee populations and concluded that, as people dismantle habitat, as climate change destabilizes food, and as disturbance inhibits normal behavior, ape populations are declining at 2.5 to 6 percent annually. "Considerable effort is urgently needed to protect these populations," the scientists asserted as they pleaded for plans to protect chimpanzees' "cultural heritage."

"It's not just the loss of 'chimps,'" Cat repeats. "I find terrifying the possibility of losing each population's unique culture. That's permanent." Culture isn't just a boutique concern. The keepers of cultural knowledge allow populations to survive in their habitats. Cultures and habitats, both, are needed; both must be rescued from the current ransacking of the planet. Cultural diversity is raw material for resilience and adaptability to change. And change is accelerating.

Free-living chimpanzees still total about four hundred thousand; a species viable enough to make conservation highly worthwhile. But the four regional races have very different numbers. The form ranging from central to eastern Africa, such as those here in Uganda, still number about two hundred thousand. But the race living in Nigeria and Cameroon may be down to only six thousand individuals.

Yet Cat also believes that there will be free-living chimps in African forests for the foreseeable centuries. She finds a package of hope in protected areas and World Heritage designations of forests, a continual on-the-ground presence by scientists and rangers, and support from tourists who value the chimpanzees — and show it by bringing money. Cat predicts a mosaic of disappearance and endurance. "And that's about as optimistic as I can be," she concludes.

Perhaps that will be enough to portal them through peak humans, the densest and perhaps most brutal swarm of us, and to buy them a real million-year future. Les Kaufman of Boston University — we met him previously, in our discussion of cichlid fishes and cultural evolution — once memorably wrote to me, "I have learned from studying species entering the extinction vortex that it's actually hard to make something go extinct. There is a protracted time during which conservation efforts can pay off." However, he added a note of caution: "The tragedy is that we seldom take advantage of this fact, though a deep love of nature exists in many people." When we do, it works; conservation efforts have reversed near-certain oblivion for various birds, mammals from rodents to whales, and dozens of others.

When I was in grad school, in the mid-1980s, I took a special trip, part pilgrimage and part requiem, to see a free-living California condor when only *six* remained in the wild. The condors were being fatally poisoned by lead, and those last six were scheduled to be captured and added to the two dozen living in captivity, making the species extinct in the wild in order to save it from oblivion due to an environment turned toxic. More than three decades later, as I was finishing this book, the recovery program's one thousandth condor egg had hatched; nearly three hundred now fly free. No California condor would yet corkscrew into the sky had Congress not passed the Endangered Species Act of 1973.

Some of the turnarounds have been triumphs for large, much-loved species: Roughly 80 percent of marine mammals and 75 percent of sea turtles that attained the protections of the U.S. Endangered Species Act have increased significantly. The Endangered Species Act has helped inspire other countries to succeed in other noteworthy turnarounds. The black rhino had declined 98 percent from its 1960 numbers because of poaching. Despite the loss of one subspecies, aggressive conservation has allowed rhinoceros numbers to rise to about five thousand. The rarest of nine giraffe subspecies was down to fifty individuals remaining in Niger; its numbers have multiplied to about four hundred. Gray whales were hunted to extinction in the Atlantic and nearly so in the Pacific, but they've recovered spectacularly along North America's western coast and are often seen from the shore, from Mexico to Alaska. Many people, including myself, have traveled to the whales' Baja birthing lagoons, watched in amazement as a wild whale has come to the side of the boat, and petted the whale's baby. These are just a few examples of what we can do. The cheering fact is: when people want it to, saving species works. The animals just need room to live and to be left in peace to make their own choices.

Deep in Sonso territory, we come to a picturesque rock-rimmed pool of water. Ten or so chimpanzees repose in nearby shadows, dark shapes in dark shade. Cat effortlessly takes attendance, noting each chimp by name.

Irene emerges from the shadows with her six-month-old infant, Ishe, riding her shoulders. Because Ishe is tiny and vulnerable, Irene greets *everyone.* Formalities concluded, mother and baby array themselves in dappled sunlight on a fallen tree, just uphill from the water holes. It feels to me, here in Sonso, that a pendulum has swung toward peace.

I lower myself to the ground, placing my knuckles where innumerable chimps over incalculable time have placed theirs. In the softening light of late afternoon, rambunctious babies nurse and play, hug their mums, swing on vines, and chase each other. They provoke wrestling from friendly adults by tapping them or grabbing a foot.

Little Muhumuza jumps off his mother Melissa's lap, swings around a sapling, goes back to her, hops on, hops off, repeats, slides into the leaves to sit beside her. He scampers over to some bigger kids who are chasing each other. But tiny Muhumuza soon quickly scampers back to the safety of his mother's arms.

For two hours, we live among this loveliness.

Cat's right, of course: their life is "like" *many* things, lived in many ways. When it's not peaceful, the dustups and the screaming, the displays, the disturbances — all of that unrest colors your thoughts about them. But

I am absorbing the realization that, on an hour-by-hour basis, life for them is peaceful, *most* of the time. Scientific data shows that chimpanzees spend the vast majority of their time — roughly 99 percent of their natural lives — in peace. The same social system that costs them in tensions and upheavals pays dividends of community and relationships. This is the balance they forge with what skills they have, within the limitations of who they are and the mixed consequences of male ambitions.

Other species living within their own community networks of close family and familiar friends — most notably bonobos, sperm whales, elephants, and a few others — have found more consistent peace. But chimpanzees are who they are. And we are what we are. We all have our limitations *and* our superlatives. Temporarily and imperfectly, but for the vast majority of their lives and times, the chimpanzees connect amicably and quell their worst impulses.

And our own human ability to quell our worst impulses and to connect — even imperfectly, even temporarily, but occasionally radiantly — is perhaps the best balm for a strange ape trapped within a small hourglass of time, who seesaws between past and future as if seeking some eternally elusive level pause, struggling through the ups and downs to remember and to look ahead, obsessed with the faces of our kind that burn with rage, with fear, and with such incandescent beauties. The chimps are the best they can be. The question for us: Are we? The chimps ask no more of themselves. We must ask no less. Look into the mirror. Recognize our shared humanity, our shared limitations, shared dilemmas — our shared gift Chimpanzees tap into their better nature 99 percent of the time. Let their success be a challenge to us.

When the resting is finished and the chimpanzees begin filtering away, everyone goes to the right except Simon. He gazes down the trail in the direction his comrades have gone. Then he walks in the opposite direction, marching to his own drum, alone. He picks up his pace. He calls. Being alone, he's a bit nervous. No backup if there's trouble. He continues south, south, south. He's going somewhere, to someone he has in mind.

EPILOGUE

A sperm whale learns who she will be journeying with, a macaw casts a covetous eye on a beautiful neighbor, a chimpanzee learns to pay to play. Culture creates vast stores of unprogrammed, unplanned knowledge. The whole world speaks, sings, and shares the codes.

On the ground, this is merely very interesting. But zoom out to the big picture. Life on Earth, an infinitesimal fragment of all cosmic matter and energy, is the universe becoming aware of itself. And culture is Life adjusting and responding, over ages and in real time, to the corner of this galaxy in which it finds itself. The magic and mystery of that aware, flexible responding, manifest in everything from a singing sparrow to the Hubble Space Telescope, sends shivers.

Life is our tiny part of the universe taking charge of directing its own destiny. And Life has, in the most real ways, chosen random acts of beauty. Not all life, not all the time, but over time, over the hundreds of millions of years of this wondrous journey, this has been a trend: Life *has created* a perceptual capacity that is felt as *beauty,* and then has sought more and more of what is beautiful. Life prefers what is lovely and sees *as lovely* what it prefers. Life has chosen to see itself, and our fleck of the sky — as beautiful. That realization is so stunning it can quite take our breath away. It makes our living world not just a cosmic accident of physics and chemistry, but a miracle.

Miraculous doesn't mean safe.

While I was writing this book, the little blue macaw known as Spix's macaw (*Cyanopsitta spixii*) was declared extinct in the wild. The scientists wrote this eulogy:

Last known from near the rio São Francisco in north Bahia, Brazil, where only three birds remained in 1985–1986, and these were captured for trade in 1987 and 1988. However, a single male, paired with a female Blue-winged Macaw *Propyrrhura maracana,* was discovered at the site in July 1990, and survived until the end of 2000. There have been no subsequent records of wild birds, despite searches and fieldworker presence in the area.

A couple hundred of these macaws remaining in captivity give hope, and now will come the arduous effort to rewild captive-raised birds. But where can they go, when the scientists noted that "loss of gallery woodland and trapping for the cagebird trade likely drove declines"?

Many skills of living must be learned from elders who learned from *their* elders. If that chain breaks, survival drops and the living world shrivels just a little bit more. Recovery, never assured, requires more time and treasure and becomes a more tenuous prospect.

Beings who've succeeded on Earth for millions of years don't seek, and should not require, our approval. They belong as well as do we. We do ourselves no favors by asking whether their existence is worth our while. We are hardly in a position to judge, hurtling and lurching along as we are with no goal, no plan except: bigger, faster, more.

If we had the courage to be honest about it, we would have to admit that whales and birds and apes and all the rest live fully up to everything of which they are capable. And we, regrettably, fall short of doing that. For them, to be is enough. For us in the isolating alienation of our tidal retreat from Life, nothing is enough. It is strange how dissatisfied we insist on being, when there is so much of the world to know and love.

Who are we journeying with? I've tried to peek with you into the vast answer. Where they live, in their lands and waters in the remaining reaches of the original world, they carry on vividly. As we do, they exert themselves to stay alive and to keep their babies alive. We are in many ways not so different. We are all kin, living together in this miracle.

So a newer question looms: Will we let them continue to exist or will we finalize their annihilation? That's our stark choice.

It used to be that animals did not need us. Now they do. Unless we value their existence, the modern tide will engulf and obliterate them. Might we reconnect?

No religion has ever preached that our role is to leave less for those who'll come after us. No wisdom tradition grants a generation permission to deplete the world and drive it toward ruin. We are taught instead that we must pilot the ark to safety. We must help them outlast the crisis of our time. We can never save the world, though we can spoil it. Life is a relay race, our task merely to pass the torch, along with a world that is at least as alive as the one we inherited, and perhaps a bit better. Caring that they'll exist after we are gone is a moral matter.

But there are practical concerns for humans, too, and for the descendants of our own kind known as our children and their children. And our choice for them is, if anything, even starker. The things that threaten whole communities of other species also threaten us. How could they not? Degradation of land, exhaustion of soil, sickening pollution of water and air and food, the ways we have dismissively destabilized the life support on this planet we casually call the world —. By these things we pose a threat not just to ourselves and others now, but to those who will be coming next and down the line. When species upon species become endangered it means deeply systemic incompatibilities have broken out into symptomatic illnesses. Upon our rocky life raft in space, living things anchor what is beautiful. As they dwindle, the beautiful drains away. As animals and plants lose their grip on existence, we lose the beauty of our own. Beauty is a simple crib-note for all that matters. A thing is right when it adds, or aids, what it beautiful. Parks and refuges, necessary now but not sufficient, are merely the flip side of widespread demolition; they are like preserving Mona Lisa's eyes while pulping the rest of Da Vinci's masterpiece for its fibers, then congratulating ourselves for our foresight. There can be no substitute for expanding the true, deep, broad coexistence that is required. Averting the demise and further degradation of beauty is crucial for human dignity, for human sanity.

Can we evolve a culture for a beautiful future on Earth? Only humans can ask that question. Only humans need to. And everything that means anything depends on our answer.

NOTES

REALM ONE. RAISING FAMILIES: SPERM WHALES

Families, One

The first description of their clicks: Worthington, L. V., and W. E. Schevill. 1957. "Underwater Sounds Heard from Sperm Whales." *Nature* 180: 291.

"a 'pop' so loud": Ellis, *Great Sperm Whale,* p. 4.

the oldest known pregnant female: Whitehead, *Sperm Whales,* p. 123.

three to as many as forty clicks: Oliveira, C., et al. 2016. "Sperm Whale Codas May Encode Individuality as Well as Clan Identity." *Journal of the Acoustical Society of America* 139: 2860–69.

"a particularly heavy burst": Whitehead, *Voyage to the Whales,* p. 151.

"clearly cultural": Norris, K. S., and C. R. Schilt. 1988. "Cooperative Societies in Three-Dimensional Space: On the Origins of Aggregations, Flocks and Schools, with Special Reference to Dolphins and Fish." *Ethology and Sociobiology* 9: 149–79.

pilot whales: Van Cise, A. M., et al. 2018. "Song of My People: Dialect Differences Among Sympatric Social Groups of Short-Finned Pilot Whales in Hawai'i." *Behavioral Ecology and Sociobiology* 72: 193.

"no parallel": Rendell, L., and H. Whitehead. 2001. "Culture in Whales and Dolphins." *Journal of Behavioral and Brain Science* 24: 309–82.

voices are symbols *of identity:* Ribeiro, S., et al. 2007. "Symbols Are Not Uniquely Human." *Biosystems* 90: 263–72.

The need for reliable babysitters: Gero, S., and H. Whitehead. 2007. "Suckling Behavior in Sperm Whale Calves." *Marine Mammal Science* 23: 398–413.

"mother culture" and *Babies oft n trundle:* Whitehead and Rendell, *Cultural Lives of Whales and Dolphins,* pp. 126–61.

more than five thousand miles: Rasmussen, K., et al. 2007 "Southern Hemisphere Humpback Whales Wintering off Central America: Insights from Water Temperatures into the Longest Mammalian Migration." *Biology*

Letters 3: 302–5.

females who'd been tagged together and *constant companions:* Whitehead, *Sperm Whales,* p. 222.

"with the most unceasing care and fondness": Beale, *Natural History of the Sperm Whale.*

most *squid they eat:* Madsen, P. T., et al. 2007. "Clicking for Calamari: Toothed Whales Can Echolocate Squid *Loligo pealeii.*" *Aquatic Biology* 1: 141. See also: Kawakami, T. 1980. "A Review of Sperm Whale Food." *Scientific Reports of the Whales Research Institute* 32: 199–218.

four thousand squid mandibles: Clarke, M. R. 1962. "Stomach Contents of a Sperm Whale Caught off Madeira in 1959." *Norsk Hvalfangst-tidende,* 173–91, cited in Ellis, *Great Sperm Whale.*

Researchers have determined: Madsen et al., "Clicking for Calamari," 141–50.

as many as ten thousand: Whitehead and Rendell, *Cultural Lives of Whales and Dolphins,* p. 156.

Sperm whale clans: Rendell, L., and H. Whitehead. 2003. "Vocal Clans in Sperm Whales." *Proceedings of the Royal Society B* 270: 225–31.

"the second inheritance": Whiten, A. 2017. "A Second Inheritance System: The Extension of Biology Through Culture." *Interface Focus.* Online.

and both evolve: Mesoudi, A. 2017. "Pursuing Darwin's Curious Parallel: Prospects for a Science of Cultural Evolution." *Proceedings of the National Academy of Sciences* 114: 7853–60.

"Culture, we believe": Whitehead and Rendell, *Cultural Lives of Whales and Dolphins,* p. 17.

Families, Three

opposite sides of the Pacific and, below, *"culturally transmitted behavior":* Gaëtan, R., et al. 2018. "Cultural Transmission of Fine-Scale Fidelity to Feeding Sites May Shape Humpback Whale Genetic Diversity in Russian Pacific Waters." *Journal of Heredity* 109: 724–34. See also: Urban, J., et al. 2000. "Migratory Destinations of Humpback Whales Wintering in the Mexican Pacific." *Journal of Cetacean Research and Management* 2: 101–10.

concentrate in Hawaiian waters: Herman, L. M. 1979. "Humpback Whales in Hawaiian Waters: A Study in Historical Ecology." *Pacific Science* 33: 1–15.

beluga whales: Anon. 2018. "Like Human Societies, Whales Value Culture and Family Ties." *Science Daily.* Online.

Right whales: Mate, B. R., et al. 1997. "Satellite-Monitored Movements of the Northern Right Whale." *Journal of Wildlife Management* 61: 1393–405.

"Songs of Humpback Whales": Payne, R. S., and S. McVay. 1971. "Songs of Humpback Whales." *Science* 173: 585–97.

"unparalleled in any other nonhuman": Garland, E. C., et al. 2011. "Dynamic Horizontal Cultural Transmission of Humpback Whale Song at the Ocean Basin Scale." *Current Biology* 21: 687–91.

"Normally you don't hear": Rothenberg, *Thousand Mile Song.*

ingredients of margarine: Whitehead and Rendell, *Cultural Lives of Whales and Dolphins,* p. 76.

Families, Four

kittens: Thornton, A., and N. J. Raihani. 2008. "The Evolution of Teaching." *Animal Behaviour* 75: 1823–36.

One definition of culture: Holzhaider, J. C., et al. 2010. "Social Learning in New Caledonian Crows." *Learning and Behavior* 38: 206.

"the way we do things": McGrew, *Cultured Chimpanzee.*

whacking the water surface: Allen, J., et al. 2013. "Network-Based Diffusion Analysis Reveals Cultural Transmission of Lobtail Feeding in Humpback Whales." *Science* 340: 485–88.

longlines that are miles long: Schakner, Z. A. 2014. "Using Models of Social Transmission to Examine the Spread of Longline Depredation Behavior Among Sperm Whales in the Gulf of Alaska." *PLOS One* 9: 109079.

killer whale (orca) to imitate: Abramson, J. Z., et al. 2013. "Experimental Evidence for Action Imitation in Killer Whales (*Orcinus orca*)." *Animal Cognition* 16: 11–22.

more than eighty coda types: Whitehead, *Sperm Whales,* p. 290.

3,190 meters deep: Ibid., p. 81.

Families, Five

"No sooner did the herd": Melville, *Moby-Dick,* ch. 87.

Jacques Cousteau speculated: Cousteau and Dumas, *Silent World,* pp. 206–7.

Around thirty-five million years: Lindberg, D. R., and N. D. Pyenson. 2007. "Things That Go Bump in the Night: Evolutionary Interactions Between Cephalopods and Cetaceans in the Tertiary." *Lethaia* 40: 335–43.

"male-male aggression": Carrier, D. R., et al. 2002. "The Face That Sank the *Essex:* Potential Function of the Spermaceti Organ in Aggression." *Journal of Experimental Biology* 205: 1755–63.

about twenty times as large: Whitehead, *Sperm Whales,* p. 318.

To generate their sonar: Cranford, T. W. 2000.

"In Search of Impulse Sound Sources in Odontocetes." In *Hearing by Whales and Dolphins* (Springer Handbook of Auditory Research series), ed. W.W.L. Au, A. N. Popper, and R. R. Fay (New York: Springer-Verlag).

Richard Ellis commented: Ellis, *Great Sperm Whale,* p. 113.

the "acoustic funnel": Yamato, M., and M. D. Pyenson. 2015. "Early Development and Orientation of the Acoustic Funnel Provides Insight into the Evolution of Sound Reception Pathways in Cetaceans." *PLOS One* 10: e0118582.

Some blind people: Norman, L. J., and L. Thaler. 2019. "Retinotopic-like Maps of Spatial Sound in Primary 'Visual' Cortex of Blind Human Echolocators." *Proceedings of the Royal Academy B.* Online doi:10.1098/rspb.2019.1910. See also: Servick, K. 2019. "Echolocation in Blind People Reveals the Brain's Adaptive Powers." *Science.* Online. doi:10.1126/science.aaz7018.

a billion clicks: Whitehead and Rendell, *Cultural Lives of Whales and Dolphins,* p. 147.

sperm whales can hear these clangs: Whitehead, *Sperm Whales,* p. 144.

"I had expected these huge males": Gordon, J. 1998. *Sperm Whales* (Grantown-on-Spey, Scotland: Colin Baxter), pp. 22–25.

Off the Galápagos Islands: Whitehead and Rendell, *Cultural Lives of Whales and Dolphins,* pp. 152–54.

culture leads, and genes follow: Ibid. See also: Danchin, E., et al. 2004. "Public Information: From Nosy Neighbors to Cultural Evolution." *Science* 305: 487–91. And also: Aplin, L. M. 2019. "Culture and Cultural Evolution in Birds: A Review of the Evidence." *Animal Behaviour* 147: 179–87. And also: Foote, A. D., et al. 2016. "Genome-Culture Coevolution Promotes Rapid Divergence of Killer Whale Ecotypes." *Nature Communications* 7: 11693.

dolphin groups specialize: Gazda, S. K., et al. 2005. "A Division of Labour with Role Specialization in Group-Hunting Bottlenose Dolphins (*Tursiops truncatus*) off Cedar Key, Florida." *Proceedings of the Royal Society of London B* 272: 135–40.

In the Adriatic Sea: Genov, T. 2019. "Behavioural and Temporal Partitioning of Dolphin Social Groups in the Northern Adriatic Sea." *Marine Biology* 166: 11.

arrival of shrimp trawlers: Chilvers, B. L., et al. 2003. "Influence of Trawling on the Behavior and Spatial Distribution of Indo-Pacific Bottlenose Dolphins (*Tursiops aduncus*) in Moreton Bay, Australia." *Canadian Journal of Zoology* 81: 1947–55.

Families, Seven

Mocha Dick: J. N. Reynolds. "Mocha Dick; or, The White Whale of the Pacific: A Leaf from a Manuscript Journal." *The Knickerbocker, or New-York Monthly Magazine* 13, no. 5 (May 1839): pp. 377–92; http://bit.ly/2xUwEsp.

Owen Chase: Chase's account is quoted on pp. 16–17 of Ellis's *Great Sperm Whale.*

"at a loss to assign a motive": Ellis, *Great Sperm Whale,* p. 22.

Families, Eight

Whale experts Bob Pitman: Pitman, R. L., et al. 2001. "Killer Whale Predation on Sperm Whales: Observations and Implications." *Marine Mammal Science* 17: 494–507.

In the northern Gulf: Whitt, A., et al. 2015. "First Report of Killer Whales Harassing Sperm Whales in the Gulf of Mexico." *Aquatic Mammals* 41: 252–55.

fending off harassing pilot whales: Weller, David, et al. 1996. "Observations of Interaction Between Sperm Whales and Short-Finned Pilot Whales in the Gulf of Mexico." *Marine Mammal Science* 12: 588–94.

"the apparent helplessness": Pitman et al., "Killer Whale Predation on Sperm Whales," 494–507.

sperm whale groups vary from three: Whitehead, *Sperm Whales,* p. 232.

"Owing to the unwearied": Moby-Dick, ch. 87.

"We glided between": Ibid.

"They lie in the uterus": Beale, *Natural History of the Sperm Whale,* pp. 52, 126.

"The unborn whale": Moby-Dick, ch. 87.

"I have seen in one school": Beale, *Natural History of the Sperm Whale,* p. 51.

"In former years": Moby-Dick, ch. 105.

"as if numerous nations": Ibid., ch. 87.

"a few individuals in America": Beale, *Natural History of the Sperm Whale,* pp. 141–42.

"at a vast expense": Ibid., 148; the following information, including that about the *Syren,* is from p. 150.

Comte Bernard-Germain de Lacépède: as quoted in Ellis, *Great Sperm Whale,* p. 314.

"Owing to the almost omniscient" and *"We account the whale immortal":* Moby-Dick, ch. 105.

Families, Nine

Ten regional whale populations and *For all whales everywhere:* Balance, L. T. 2014. "Whaling: Past, Present, and Future." Online.

From the 1700s; sixty years; the 1960s; three million; and *beat them all:* Cressey, D. 2015. "World's Whaling Slaughter Tallied." *Nature* 519: 140–41.

"blue whale unit": "Blue Whale Unit Limit,"

from "Chairman's Report of the Fifteenth Meeting," 1963; http://luna.pos.to/whale/iwc_chair63_8.html.

In the sailing-whaling era: Whitehead, *Sperm Whales,* p. 20.

Between 1964: Ellis, *Great Sperm Whale,* pp. 272–93.

killed 48,702 and 3,212 southern right whales: Clapham, P., et al. 2009. "Catches of Humpback Whales, *Megaptera novaeangliae,* by the Soviet Union and Other Nations in the Southern Ocean, 1947–1973." *Marine Fisheries Review* 71: 39–43.

Peter Matthiessen saw: Matthiessen, *Blue Meridians,* pp. 4–18.

A. A. Berzin warned: Ivashchenko, Y. V., et al. 2008. "The Truth About Soviet Whaling: A Memoir, by A. A. Berzin [translated by Y. V. Ivashchenko]." *Marine Fisheries Review* 70: 1–59.

In 1974: Ellis, *Great Sperm Whale,* p 272; Ellis, *Great Sperm Whale,* p. 284; Whaling Commission quotas and Soviet whaling: Ellis, *Great Sperm Whale,* pp. 290–93.

votes for a moratorium: Ellis, *Great Sperm Whale,* pp. 283–303.

DNA of seven hundred pieces: Baker, C. S., et al. 2000. "Predicted Decline of Protected Whales Based on Molecular Genetic Monitoring of Japanese and Korean Markets." *Proceedings of the Royal Society B* 267:

1191–99.

"research" while doing no science: BBC News. 2015. "Japan to Resume Whaling in Antarctic Despite Court Ruling." Online.

Proposals to kill more whales: Denyer, S., and I. Kasiwagi. 2018. "Japan to Leave International Whaling Commission, Resume Commercial Hunting." *Washington Post.* Online.

"It's as if intelligent": Ellis, *Great Sperm Whale,* p. 310.

Often they poop immediately: Doughty, C. E., et al. 2015. "Global Nutrient Transport in a World of Giants." *Proceedings of the National Academy of Sciences.* Online.

Hal Whitehead has written: Angier, N. 2010. "Save a Whale, Save a Soul, Goes the Cry." *New York Times,* June 26.

Roger Payne and persistent toxic chemicals: NPR's *Living on Earth,* online. See also: Ferber, D. 2005. "Sperm Whales Bear Testimony to Worldwide Pollution." *Science* 309: 1166b. And: Sonne, C., et al. 2018. "Pollution Threatens Toothed Whales." *Science* 361: 1208. And: Meyer, W. K., et al. 2018. "Ancient Convergent Losses of Paraoxonase 1 Yield Potential Risks for Modern Marine Mammals." *Science* 361: 591–94.

"The remnant bands": Matthiessen, *Blue Meridians,* p. 6.

Families, Ten

Seventy-five ocean scientists: Author's files.

A multiyear study of whale stress-hormone levels: Rolland, R. M., et al. 2012. "Evidence That Ship Noise Increases Stress in Right Whales." *Proceedings of the Royal Society B* 279: 2363–68.

By 2000 and *for the next decade:* Cantor, M., et al. 2016. "Cultural Turnover Among Galápagos Sperm Whales." *Royal Society Open Science.* Online.

Toklat pack: Dutcher, J., and J. Dutcher. 2018. *The Wisdom of Wolves* (Washington, DC: National Geographic), pp. 132–33.

"The Sperm Whale, with a maximum rate": Red List: Sperm Whale; http://www.iucnredlist.org/details/41755/0.

Families, Eleven

"The whales were hanging": Whitehead, *Voyage to the Whales,* p. 115.

"I heard a thunderous crack": Paraphrased from: Nestor, J., 2016. "A Conversation with Whales." *New York Times,* April 16. Online.

REALM TWO. CREATING BEAUTY: SCARLET MACAWS

Beauty, One

three hundred million years ago: Zhang, G., et al. 2014. "Comparative Genomics Reveals Insights into Avian Genome Evolution and Adaptation." *Science* 346: 1311–20.

"the humans of the bird world": Klein, J. 2018. "The Genes That Make Parrots into the Humans of the Bird World." *New York Times.* Online.

parrot's brain and brain stem: Iwaniuk, A. N., et al. 2005. "Interspecific Allometry of the Brain and Brain Regions in Parrots (Psittaciformes): Comparisons with Other Birds and Primates." *Brain, Behavior and Evolution* 65: 40–59.

validity of the mirror test: Personal communication with one of the researchers. See Kohda, M., et al. 2019. "Cleaner Wrasse Pass the Mark Test. What Are the Implications for Consciousness and Self-Awareness Testing in Animals?" *PLOS Biology* 17: e3000021.

Parrots have been and *the first apes:* Emery, N. J. 2004. "Are Corvids 'Feathered Apes'? Cognitive Evolution in Crows, Jays, Rooks and Jackdaws." In *Comparative Analysis of Minds,* ed. S. Watanabe (Tokyo: Keio University Press), pp. 181–213.

"thought until not long ago": Seed, A., et al. 2009. "Intelligence in Corvids and Apes: A Case of Convergent Evolution?" *Ethology* 115: 401–20.

parrots and birds of the crow family: Ibid.

New Caledonia crows: Holzhaider, J. C., et al. 2010. "Social Learning in New Caledonian Crows." *Learning and Behavior* 38: 206–17.

"multiple traditions": Whiten, A., and C. P. van Schaik. 2007. "The Evolution of Animal 'Cultures' and Social Intelligence." *Philosophical Transactions of the Royal Society B* 362: 603–20.

"the birds performed": Kabadayi, C., and M. Osvath. 2017. "Ravens Parallel Great Apes in Flexible Planning for Tool-Use and Bartering." *Science* 357: 202–4.

ravens can follow a human's gaze: Schloegl, C., et al. 2007. "Gaze Following in Common Ravens, *Corvus corax:* Ontogeny and Habituation." *Animal Behaviour* 74: 769–78.

delayed gratification: Krasheninnikova, A., et al. 2018. "Economic Decision-Making in Parrots." *Scientific Reports* 8, article 12537.

two separate pinnacles: Emery, N. J., and N. S. Clayton. 2004. "The Mentality of Crows: Convergent Evolution of Intelligence in Corvids and Apes." *Science* 306: 1903–7.

an African grey parrot named Griffin: Clements, K. A., et al. 2018. "Initial Evidence for Probabilistic Reasoning in a Grey Parrot." *Journal of Comparative Psychology* 132: 166–77.

parrots can generalize: Pepperberg, I. M., and K. Nakayama. 2016. "Robust Representation of Shape in a Grey Parrot." *Cognition* 153: 146–60.

some species of crows, scrub jays, and ravens: Emery and Clayton, "The Mentality of Crows," 1903–7.

acorn-storing sites: Personal conversation with Matthew Fuirst, Stony Brook University.

individuals differ in personality: DiRienzo, N., and A. Hedrick. 2014. "Animal Personalities and Their Implications for Complex Signaling." *Current Zoology* 60: 381–86.

Female parakeets: Chen, J., et al. 2019. "Problem-Solving Males Become More Attractive to Female Budgerigars." *Science* 11: 166–67.

chickadee-like great tits: Estok, P., et al. 2010. "Great Tits Search for, Capture, Kill and Eat Hibernating Bats." *Biology Letters* 6: 59–62.

herring gulls exploit: Fuirst, M., et al. Manuscript. "Effects of Urbanization on the Foraging Ecology and Microbiota of a Generalist Seabird."

sea otters learn a foraging: Estes, J. A., et al. 2003. "Individual Variation in Prey Selection by Sea Otters: Patterns, Causes and Implications." *Journal of Animal Ecology* 72:

144–55.

oystercatchers: Norton-Griffiths, M. 1967. "Some Ecological Aspects of the Feeding Behaviour of the Oystercatcher *Haematopus ostralegus* on the Edible Mussel *Mytilus edulis.*" *Ibis.* Online.

only one in Minnesota: Kraker, D. 2018. "The Secret Fishing Habits of Northwoods' Wolves." NPR.org. Online.

One morning in 1921: Aplin, L. M., et al. 2013. "Milk Bottles Revisited: Social Learning and Individual Variation in the Blue Tit, *Cyanistes caeruleus.*" *Animal Behaviour* 85: 1225–32. See also: Aplin, L. M., et al. 2015. "Experimentally Induced Innovations Lead to Persistent Culture via Conformity in Wild Birds." *Nature* 518: 538–41.

motion sensors to open doors: Breitwisch, R., and M. Breitwisch. 1991. "House Sparrows Open an Automatic Door." *Wilson Bulletin* 103: 725–26. See also: Spector, D. 2014. "Smart Birds Learned How to Operate Automatic Doors." *Business Insider.* Online.

Urban sparrows and finches: Suárez-Rodriguez, M., et al. 2012. "Incorporation of Cigarette Butts into Nests Reduces Nest Ectoparasite Load in Urban Birds: New Ingredients for an Old Recipe?" *Biology Letters* 9: 20120931.

world's nine thousand or so: Waters, H. 2016.

"New Study Doubles the World's Number of Bird Species by Redefining 'Species.'" *Audubon.* Online.

Dupont's lark: Laiolo, P., and J. L. Tella. 2007. "Erosion of Animal Cultures in Fragmented Landscapes." *Frontiers in Ecology and the Environment* 5: 68–72.

orange-billed sparrow and *cultural diversity:* Hart, P. J., et al. 2018. "Birdsong Characteristics Are Related to Fragment Size in a Neotropical Forest." *Animal Behaviour* 137. Online.

Beauty, Two

Dipteryx: Brightsmith, D. J. 2005. "Parrot Nesting in Southeastern Peru: Seasonal Patterns and Keystone Trees." *Wilson Bulletin* 117: 296–305.

Worldwide, nearly a third of parrots: Berkunsky, I., et al. 2017. "Current Threats Faced by Neotropical Parrot Populations." *Biological Conservation.* Online.

In a famous series of studies: Nijhuis, M. 2008. "Friend or Foe? Crows Never Forget a Face, It Seems." *New York Times,* August 25. Online.

nestlings develop individually unique calls: Berg, K. S., et al. 2011. "Vertical Transmission of Learned Signatures in a Wild Parrot." *Proceedings of the Royal Society B* 279: 585–91.

budgerigars: Hile, A. G., and G. F. Striedter. 2000. "Call Convergence Within Groups of Female Budgerigars (*Melopsittacus undulatus*)." *Ethology* 106: 1105–14.

chickadees: Nowicki, S. 1983. "Flock-Specific Recognition of Chickadee Calls." *Behavioral Ecology and Sociobiology* 12: 317–20.

fruit bats: Prat, Y., et al. 2017. "Crowd Vocal Learning Induces Vocal Dialects in Bats: Playback of Conspecifics Shapes Fundamental Frequency Usage by Pups." *PLOS Biology* 15: e2002556.

learn the dialects of the crowds they're in: Aplin, L. M. 2019. "Culture and Cultural Evolution in Birds: A Review of the Evidence." *Animal Behaviour* 147: 179–87.

Ravens: Massen, J.J.M., et al. 2014. "Ravens Notice Dominance Reversals Among Conspecifics Within and Outside Their Social Group." *Nature Communications* 5: 3679. Online.

In Brazil, some dolphins: Simões-Lopes, P. C., et al. 1998. "Dolphin Interactions with the Mullet Artisanal Fishing on Southern Brazil: A Qualitative and Quantitative Approach." *Revista Brasileira de Zoologia* 15: 709–26. See also: Daura-Jorge, F. G., et al. 2012. "The Structure of a Bottlenose Dolphin Society Is Coupled to a Unique Foraging Cooperation with Artisanal Fishermen." *Biology Letters* 8: 702–5. See also: Machado, A.M.S., et al. 2019. "Homophily Around Specialized Foraging Underlies Dolphin Social Preferences." *Biology Letters.* Online.

The ones who do sound different: Romeau, B., et al. 2017. "Bottlenose Dolphins That Forage with Artisanal Fishermen Whistle Differently." *Ethology* 123: 906–15.

reintroduce thick-billed parrots: Burger, *Parrot Who Owns Me,* p. 139.

storks, vultures, eagles, and hawks: Meyberg, B. 2017. "Orientation of Native Versus Translocated Juvenile Lesser Spotted Eagles (*Clanga pomarina*) on the First Autumn Migration." *Journal of Experimental Biology* 220: 2765–76.

"a potentially very significant realm": Whiten, A. 2017. "A Second Inheritance System: The Extension of Biology Through Culture." *Interface Focus.* Online.

moose, bison, deer: Festa-Bianchet, M. 2018. "Learning to Migrate." *Science* 361: 972. See also: Jesmer, B. R., et al. 2018. "Is Ungulate Migration Culturally Transmitted? Evidence of Social Learning from Translocated Animals." *Science* 361: 1023–25.

guppies, bluehead wrasses, and French grunts: Warner, R. W. 1988. "Traditionality of Mating-Site Preferences in a Coral Reef

Fish." *Nature* 335: 719–21.

former pets: Brightsmith, D., et al. 2005. "The Use of Hand-Raised Psittacines for Reintroduction: A Case Study of Scarlet Macaws (*Ara macao*) in Peru and Costa Rica." *Biological Conservation* 121: 465–72.

Beauty, Three

graphs called sonograms: Berg, K. S., et al. 2011. "Contact Calls Are Used for Individual Mate Recognition in Free-Ranging Green-Rumped Parrotlets." *Animal Behaviour* 81: 241–48.

a "core" and a "shell": Chakraborty, M., et al. 2016. "Core and Shell Song Systems Unique to the Parrot Brain." *PLOS One* 10: e0118496.

Darwin suspected that birds pass songs along: Darwin, C. R. Notebook M: [Metaphysics on morals and speculations on expression (1838)]. Darwin Online, http://darwin -online.org.uk/, pp. 31–32.

"a feather magnified": Beston, H. 1928. *The Outermost House* (repr., New York: Henry Holt, 1992), p. 25.

"A mockingbird does not": Visscher, J. P. 1928. "Notes on the Nesting Habits and Songs of the Mockingbird." *Wilson Bulletin* 40: 209–16. See also: Laskey, A. 1944. "A Mockingbird Acquires His Song Repertory." *Auk* 61: 211–19.

"alter the structure of their vocalizations": Chen, Y., et al. 2016. "Mechanisms Underlying the Social Enhancement of Vocal Learning in Songbirds." *Proceedings of the National Academy of Science* 113: 6641–46.

"dialects": Aplin, L. M. 2019. "Culture and Cultural Evolution in Birds: A Review of the Evidence." *Animal Behaviour* 147: 179–87.

"Cod particularly": Anon. 2016. "What Do We Mean by 'Accents' in Animals?" BBC Newsbeat. Online. See also: Anon. 2016. " 'Talking' Cod Have Regional Accents, and You Can Listen to Them Here." *IFLScience.* Online.

Yellow-naped Amazon parrots: Berg, K. S., et al. 2011. "Vertical Transmission of Learned Signatures in a Wild Parrot." *Proceedings of the Royal Society B* 279: 585–91. See also: Wright, T. F., et al. 2008. "Stability and Change in Vocal Dialects of the Yellow-Naped Amazon." *Animal Behaviour* 76: 1017–27.

singing male white-crowned sparrows and *indigo buntings' vocalizations:* Prum, R. 2013. "Coevolutionary Aesthetics in Human and Biotic Artworlds." *Biology and Philosophy* 28: 811–82. See also: MacDougall-Shackleton, E. A., and S.A. MacDougall-Shackleton. 2001. "Cultural and Genetic Evolution in Mountain White-Crowned

Sparrows: Song Dialects Are Associated with Population Structure." *Evolution* 55: 2568–75.

"there is no nature-nurture divide": Birkhead, *Bird Sense,* p. 41.

dopamine: Simonyan, K., et al. 2012. "Dopamine Regulation of Human Speech and Bird Song: A Critical Review." *Brain and Language* 122: 142–50. See also: Ackerman, *Genius of Birds,* pp. 151–54.

Beauty, Four

Birds watch one another: Burger, *Parrot Who Owns Me,* p. 155.

When Joanna said to Tiko: Ibid., p. 150.

Captive chimpanzees and bonobos and *At least one border collie:* Griebel, U., et al. 2016. "Developmental Plasticity and Language: A Comparative Perspective." *Topics in Cognitive Science* 8: 435–45.

specific training technique is required: Pepperberg, I. M. 2010. "Vocal Learning in Grey Parrots: A Brief Review of Perception, Production, and Cross-Species Comparisons." *Brain and Language* 115: 81–91.

For Alex, an apple was a "banberry": Griebel et al, "Developmental Plasticity and Language."

Prairie dogs' alarms: Slobodchikoff, C. N., et al. 2009. "Prairie Dog Alarm Calls Encode Labels About Predator Colors." *Animal Cog-*

nition 12: 435–39.

Hooded warblers recognize: Godard, R. 1991. "Long-Term Memory of Individual Neighbours in a Migratory Songbird." *Nature* 350: 228–29.

Various primates, crows, dogs: Kondo, N. 2012. "Crows Cross-Modally Recognize Group Members but Not Non-Group Members." *Proceedings of the Royal Society B* 279: 1937–42.

Pigeons, crows, and jackdaws: University of Lincoln. 2012. "Birds Can Recognize People's Faces and Know Their Voices." *ScienceDaily,* June 22.

"The lower animals": Darwin, *Descent of Man,* p. 39.

seem to play and fool around: Emery, N. J., and N. S. Clayton. 2015. "Do Birds Have the Capacity for Fun?" *Current Biology* 25: R16–20. Online.

Birds' and mammals' brains: Ibid.

In cowbirds, young males: West, M. J., et al. 2003. "Discovering Culture in Birds: The Role of Learning and Development." In *Animal Social Complexity: Intelligence, Culture, and Individualized Societies,* ed. F.B.M. de Waal and P. L. Tyack (Cambridge, MA: Harvard University Press), pp. 470–92.

Galápagos finches: Grant, B. R., and P. R. Grant. 2002. "Simulating Secondary Contact in Allopatric Speciation: An Empirical

Test of Premating Isolation." *Biological Journal of the Linnean Society* 76: 545–56.

African indigobirds: Balakrishnan, C. N., and M. D. Sorenson. 2006. "Song Discrimination Suggests Premating Isolation Among Sympatric Indigobird Species and Host Races." *Behavioral Ecology* 17: 473–78.

Beauty, Five

"The sight of a feather": Darwin, C. Letter to Asa Gray, April 3, 1860. *Darwin Online.*

the Duke of Argyll demanded: Darwin, *Descent of Man,* p. 442.

"With the great majority of animals": Ibid., p. 61.

bowerbirds: Prum, R. 2013. "Coevolutionary Aesthetics in Human and Biotic Artworlds." *Biology and Philosophy* 28: 811–32. See also: Madden, J. R. 2008. "Do Bowerbirds Exhibit Cultures?" *Animal Cognition* 11: 1–12. And: Ackerman, *Genius of Birds,* p. 175. And: https://www.youtube.com/watch?v=1XkPeN3AWIE.

"Males display their gorgeous plumage": Darwin, *On the Origin of Species,* p. 89.

"A great number of male animals": Ibid., p. 240.

"If female birds": Darwin, *Descent of Man,* p. 61.

Singing zebra finch: Prum. "Coevolutionary Aesthetics in Human and Biotic Artworlds," 811–32.

"The most refined beauty: Darwin, *Descent of Man,* p. 516.

Alfred Russel Wallace and Darwin on beauty: Prum, R. O. 2012. "Aesthetic Evolution by Mate Choice: Darwin's Really Dangerous Idea." *Philosophical Transactions of the Royal Society B* 367: 2253–65.

"The only way in which": Wallace, A. R. 1895. *Natural Selection and Tropical Nature,* 2nd ed. (New York: Macmillan), pp. 378–79.

R. A. Fisher offered a middle path: Fisher, R. A. 1958. *The Genetical Theory of Natural Selection* (New York: Dover Publications).

Beauty, Six

Female guppies like: Gasparini, C., et al. 2013. "Do Unattractive Friends Make You Look Better? Context-Dependent Male Mating Preferences in the Guppy." *Proceedings of the Royal Society* B. Online.

If a virgin female zebra finch: Burley, N. T. 2006. "An Eye for Detail: Selective Sexual Imprinting in Zebra Finches." *Evolution* 60: 1076–85. See also: Ackerman, *Genius of Birds,* p. 113.

fish called mollies: Witte, K., and B. Noltemeier. 2002. "The Role of Information in Mate-Choice Copying in Female Sailfin Mollies (*Poecilia latipinna*)." *Behavioral Ecology and Sociobiology* 52: 194–202.

documented in fruit flies: Danchin, E., et al.

2018. "Cultural Flies: Conformist Social Learning in Fruitflies Predicts Long-Lasting Mate-Choice Traditions." *Science* 362: 1025–30.

"How does Hen determine": Darwin, C. R. 1838–40. "Old & useless notes about the moral sense & some metaphysical points." *Darwin Online.*

called the tui, sings songs: Hill, S. D., et al. 2017. "Fighting Talk: Complex Song Elicits More Aggressive Responses in a Vocally Complex Songbird." *Ibis* 160: 257–68.

"Extravagance in nature": Ackerman, *Genius of Birds,* p. 161.

the fly Rhagoletis pomonella: Filchak, K. E., et al. 2000. "Natural Selection and Sympatric Divergence in the Apple Maggot *Rhagoletis pomonella.*" *Nature* 407: 739–42.

populations in the same region must have fractured: Rice, W. R. 1987. "Speciation Via Habitat Specialization: The Evolution of Reproductive Isolation as a Correlated Character." *Evolutionary Ecology* 1: 301–14.

specialists would start to diverge: Pfennig, D. W., et al. 2010. "Phenotypic Plasticity's Impacts on Diversification and Speciation." *Trends in Ecology and Evolution* 25: 459–67.

bluegill sunfish sorted: Ehlinger, T. J., and D. S. Wilson. 1988. "Complex Foraging Polymorphism in Bluegill Sunfish." *Pro-*

ceedings of the National Academy of Sciences 85: 1878–82.

pumpkinseed sunfish: Robinson, B. W., et al. 1993. "Ecological and Morphological Differentiation of Pumpkinseed Sunfish in Lakes Without Bluegill Sunfish." *Evolutionary Ecology* 7: 451–64.

fish called sticklebacks: Rundle, H. D., et al. 2000. "Natural Selection and Parallel Speciation in Sympatric Sticklebacks." *Science* 287: 306–8.

different species in the same lakes and *Midas cichlids:* Barluenga, M., et al. 2006. "Sympatric Speciation in Nicaraguan Crater Lake Cichlid Fish." *Nature* 439: 719–23.

"specialization is widespread but underappreciated": Bolnick, D. I., et al. 2003. "The Ecology of Individuals: Incidence and Implications of Individual Specialization." *American Naturalist* 161: 1–28.

experiments with Lake Victoria cichlid fishes: Verzijden, M. N., and C. Ten Cate. 2007. "Early Learning Influences Species Assortative Mating Preferences in Lake Victoria Cichlid Fish." *Biology Letters* 3: 134–36.

a similar experiment with birds: Hansen, B. T., et al. 2008. "Imprinted Species Recognition Lasts for Life in Free-Living Great Tits and Blue Tits." *Animal Behaviour* 75: 921–27.

Other studies also show: Danchin, E., and

R. H. Wagner. 2010. "Inclusive Heritability: Combining Genetic and Non-Genetic Information to Study Animal Behavior and Culture." *Oikos.* Online.

cultural selection: Danchin, E., et al. 2004. "Public Information: From Nosy Neighbors to Cultural Evolution." *Science* 305: 487–91.

Even the Galápagos: Grant, P. R., and B. R. Grant. 2018. "Role of Sexual Imprinting in Assortative Mating and Premating Isolation in Darwin's Finches." *Proceedings of the National Academy of Sciences* 115: E10879–E10887. See also: Verzijden, M. N., et al. 2012. "The Impact of Learning on Sexual Selection and Speciation." *Trends in Ecology and Evolution* 27: 511–19.

Beauty, Seven

"On the whole, birds appear": Darwin, *Descent of Man,* p. 466.

Prum has pointed out: Prum, R. O. 2012. "Aesthetic Evolution by Mate Choice: Darwin's Really Dangerous Idea." *Philosophical Transactions of the Royal Society B* 367: 2253–65.

"Many animals share" and *"It is not an accident":* Prum, R. 2013. "Coevolutionary Aesthetics in Human and Biotic Artworlds." *Biology and Philosophy* 28: 811–32.

Many birds see below: Birkhead, *Bird Sense,*

p. xviii. See also: Tedore, C., and D. Nilsson. 2019. "Avian UV Vision Enhances Leaf Surface Contrasts in Forest Environments." *Nature Communications* 10: 238.

"rings, bull's-eyes": Jabr, F. 2019. "How Beauty Is Making Scientists Rethink Evolution." *New York Times Magazine.* January 9.

"associated with low anxiety": Clarke, T., and A. Costall. 2008. "The Emotional Connotations of Color: A Qualitative Investigation." *Color Research and Application* 33: 406–10.

"when animals became mobile": Emmons, S. W. 2012. "The Mood of a Worm." *Science* 338: 475–76. See also: Garrison, J., et al. 2012. "Oxytocin/Vasopressin-Related Peptides Have an Ancient Role in Reproductive Behavior." *Science* 338: 540–43.

REALM THREE. ACHIEVING PEACE: CHIMPANZEES

"Chimpanzees are always new to me": Nishida, T. 1993. In *The Great Ape Project: Equality Beyond Humanity,* ed. P. Cavalieri and P. Singer, eds. (London: Fourth Estate), pp. 24–26.

Peace, One

"We've learned": Stanford, *New Chimpanzee,* p. 152.

About six million years ago: Anon. 2005. "A

Brief History of Chimps." *Nature* 437: 48–49.

Orangutans, roughly fifteen: van Schaik, C. P., et al. 2003. "Orangutan Cultures and the Evolution of Material Culture." *Science* 299: 102–5.

no new parts: Berns, G. S., et al. 2013. "Replicability and Heterogeneity of Awake Unrestrained Canine fMRI Responses." *PLOS One* 8: e81698.

Chimpanzees and bonobos share more than 99 percent: Hobaiter, C. Manuscript. "Gestural Communication in the Great Apes: Tracing the Origins of Language."

"Chimpanzees live long, interesting lives": Stanford, *New Chimpanzee,* p. 152.

mental concept *of "we":* McGrew, *Cultured Chimpanzee,* p. 149.

through generations, over centuries: Langergraber, K. E., et al. 2014. "How Old Are Chimpanzee Communities?" *Journal of Human Evolution* 69: 1–7.

Peace, Two

When a snare's noose closes: Boesch, *Real Chimpanzee,* p. 161.

Grooming's superficial effect: Reynolds, *Chimpanzees of the Budongo Forest,* pp. 111–15.

Average tenure for an alpha: Stanford, *New Chimpanzee,* p. 150.

Peace, Three

Pregnancy lasts eight months through *their limit of life span:* Stanford, *New Chimpanzee,* pp. 114–23.

In the bonobo brain: Rilling, J. K., et al. 2011. "Differences Between Chimpanzees and Bonobos in Neural Systems Supporting Social Cognition." *Social Cognitive and Affective Neuroscience* 7: 369–79.

"the most empathic brain": de Waal, *Mama's Last Hug,* p. 194.

bonobos unlock doors: Tan, J., et al. 2017. "Bonobos Respond Prosocially Toward Members of Other Groups." *Scientific Reports* 7: 14733.

threefold path to peace: Wrangham, R., and D. Peterson. 1996. *Demonic Males* (New York: Mariner Books).

"everything is peaceful": Parker, I. 2007. "Swingers." *New Yorker.* July. Online.

in offering consolation: Lindegaard, M. R., et al. 2017. "Consolation in the Aftermath of Robberies Resembles Post-Aggression Consolation in Chimpanzees. *PLOS One* 12: e0177725.

the alpha sires only about one-third: Stanford, *New Chimpanzee,* p. 119.

Males "can be tremendously friendly": Reynolds, *Chimpanzees of the Budongo Forest,* p. 124.

are obsessed with status: Stanford, *New Chim-*

panzee, p. 152.

Peace, Four

no European had more than an inkling: European perceptions and portrayals of chimpanzees: present: Anon. 2005. "A Brief History of Chimps." *Nature* 437: 48–49. See also: Boesch, *Real Chimpanzee,* pp. 111, 130.

In 1844, a missionary to Liberia: Savage, T. S., and J. Wyman. 1844. "Observations on the External Characters and Habits of the *Troglodytes niger.*" *Boston Journal of Natural History* 4: 362–86. Online.

"It has often been said": Darwin, *Descent of Man,* p. 51.

"Now we must redefine": Goodall, J. 1990. *Through a Window: My Thirty Years with the Chimpanzees of Gombe.* Boston: Houghton Mifflin.

various toolmakers: See: Safina, C. 2015. *Beyond Words* (New York: Henry Holt), pp. 195–98 and 349.

tools to obtain honey: Stanford, *New Chimpanzee,* pp. 162–63. Also: Boesch, *Real Chimpanzee,* p. 117.

"down to the ground with a sapling": Handwritten note in Events Book, Budongo Conservation Field Station.

chimpanzees of the northern Democratic Republic of the Congo: Max Planck Institute for Evolutionary Anthropology. "New Chimpanzee Culture Discovered: Study Describes Unique Behavioral Patterns of Bili-Uéré Chimpanzees in the DR Congo." *ScienceDaily,* February 25, 2019.

Chimpanzees' total tool kit: Whiten, A. 2005. "The Second Inheritance System of Chimpanzees and Humans." *Nature* 437: 52–55.

Even in adjacent communities: Pascual-Garrido, A. 2019. "Cultural Variation Between Neighbouring Communities of Chimpanzees at Gombe, Tanzania." *Scientific Reports* 9: article 8260. Online.

"to create workshops": Stanford, *New Chimpanzee,* pp. 159–60.

Sassandra-N'Zo River: McGrew, *Cultured Chimpanzee,* p. 5.

Nut-cracking chimps: Stanford, *New Chimpanzee,* p. 159. See also: Matsuzawa, T. 1991. "Nesting Cups and Metatools in Chimpanzees." *Behavioral and Brain Sciences* 14: 570–71.

mothers sometimes actually guide: Stanford, *New Chimpanzee,* p. 160.

"habitual scroungers": Biro, D., et al. 2003. "Cultural Innovation and Transmission of Tool Use in Wild Chimpanzees: Evidence from Field Experiments." *Animal Cognition* 6: 213.

Female chimpanzees learn tool use and *male youngsters get more preccupied:* Stanford,

New Chimpanzee, pp. 118–19, 164.

about forty tool-use behaviors: Whiten, "The Second Inheritance System of Chimpanzees and Humans," 52–55. See also: Hobaiter, C., et al. 2014. "Social Network Analysis Shows Direct Evidence for Social Transmission of Tool Use in Wild Chimpanzees." *PLOS Biology* 12, no. 9: e1001960.

orangutans have made wire tools: University of Vienna. 2018. "Re-inventing the Hook: Orangutans Spontaneously Bend Straight Wires into Hooks to Fish for Food"; https://m.phys.org/news/2018-11-re-inventing-orangutans-spontaneously-straight-wires.html. See also: van Schaik, C. P., et al. 2003. "Orangutan Cultures and the Evolution of Material Culture." *Science* 299: 102–5.

chimps brought into a sanctuary: Laland, K. N. 2011. "From Fish to Fashion: Experimental and Theoretical Insights into the Evolution of Culture." *Philosophical Transactions of the Royal Society B.* Online.

crop raiding nocturnally: Krief, S., et al. 2014. "Wild Chimpanzees on the Edge: Nocturnal Activities in Croplands." *PLOS One* 9: e109925.

"Distinctive tool-using traditions": Biro et al, "Cultural Innovation and Transmission of Tool Use in Wild Chimpanzees," 213.

"Chimpanzee communities resemble human cultures": Whiten, A. 2017. "A Second In-

heritance System: The Extension of Biology Through Culture." *Interface Focus.* Online.

Native Alaskan Athabascan: Yarber, Y., et al. 1980. *Henry Beatus, Sr.* (North Vancouver, BC: Hancock House).

mallard duckling who was adopted: Mandelbaum, R. F. 2019. "A Mallard Duckling Is Thriving — and Maybe Diving — Under the Care of Loon Parents." *Audubon,* July 12.

Goslings who have observed: Heyes, C. 2012. "What's Social About Social Learning?" *Journal of Comparative Psychology* 126: 193–202.

Captive dolphins copy: Whiten, A., and C. P. van Schaik. 2007. "The Evolution of Animal 'Cultures' and Social Intelligence." *Philosophical Transactions of the Royal Society B* 362: 603–20.

bumblebees were trained: Whiten, "A Second Inheritance System."

"Is there teaching?" and *chimps picked up:* Caro, T. M., and M. D. Hauser. 1992. "Is There Teaching in Nonhuman Animals?" *Quarterly Review of Biology* 67: 151–74.

mother chimps gave their children: Musgrave, S., et al. 2016. "Tool Transfers Are a Form of Teaching Among Chimpanzees." *Scientific Reports* 6: 34783. Online.

"A mother began to take a few steps": Refers to Altman as quoted in Caro and Hauser,

"Is There Teaching in Nonhuman Animals?"

When spotted dolphin mothers: Bender, C., et al. 2009. "Evidence of Teaching in Atlantic Spotted Dolphins (*Stenella frontalis*) by Mother Dolphins Foraging in the Presence of Their Calves." *Animal Cognition* 12: 43–53.

Various cats: Caro and Hauser, "Is There Teaching in Nonhuman Animals?"

killer whales help juveniles: Whitehead and Rendell, *Cultural Lives of Whales and Dolphins,* pp. 182–83.

chimps tend to become set in their ways: Marshall-Pescini, S., and A. Whiten. 2008. "Chimpanzees (*Pan troglodytes*) and the Question of Cumulative Culture: An Experimental Approach." *Animal Cognition* 11: 449–56.

a less efficient nut-cracking technique: Lamon, N. 2018. "Wild Chimpanzees Select Tool Material Based on Efficiency and Knowledge." *Proceedings of the Royal Society B.* Online.

"the opposite of intelligent": Whiten and van Schaik, "The Evolution of Animal Cultures and Social Intelligence."

She dyed half the corn: van de Waal, E. 2013. "Potent Social Learning and Conformity Shape a Wild Primate's Foraging Decisions." *Science* 340: 483–85.

a tuberculosis outbreak: Sapolsky, R. M., and

L. J. Share. 2004. "A Pacific Culture Among Wild Baboons: Its Emergence and Transmission." *PLOS Biology* 2: 534–41; e106. Online.

human children are more slavish: Whiten, "The Second Inheritance System of Chimpanzees and Humans."

Conformity: Luncz, L. V., and C. Boesch. 2014. "Tradition over Trend: Neighboring Chimpanzee Communities Maintain Differences in Cultural Behavior Despite Frequent Immigration of Adult Females." *American Journal of Primatology* 76: 649–57.

female Japanese macaque: Matsuzawa, T. 2015. "Sweet-Potato Washing Revisited." *Primates* 56: 285–87.

chimpanzees hunt differently: Hobaiter, C., et al. 2017. "Variation in Hunting Behaviour in Neighbouring Chimpanzee Communities in the Budongo Forest, Uganda." *PLOS One* 12: e0178065.

crude thrusting spears: Pruetz, J. D., and P. Bertolani. 2007. "Savanna Chimpanzees, *Pan troglodytes verus,* Hunt with Tools." *Current Biology* 17: 412–17.

Peace, Five

FOXP2: Anon. 2005. "A Brief History of Chimps." *Nature* 437: 48–49.

Gesture becomes communication: Kersken, V.,

et al. 2018. "A Gestural Repertoire of 1- to 2-Year-Old Human Children: In Search of the Ape Gestures." *Animal Cognition.* Online. See also: Hobaiter, C., and R. W. Byrne. 2014. "The Meanings of Chimpanzee Gestures." *Current Biology* 14: 1596–1600.

sixty-six different intentional gestures: Kersken et al., "A Gestural Repertoire of 1- to 2-Year-Old Human Children."

Gorillas have a total repertoire: Genty, E., et al. 2009. "Gestural Communication of the Gorilla (*Gorilla gorilla*): Repertoire, Intentionality and Possible Origins." *Animal Cognition* 12: 527–46.

Ravens gesture: Pika, S., and T. Bugnyar. 2011. "The Use of Referential Gestures in Ravens (*Corvus corax*) in the Wild." *Nature Communications* 2: 560.

Dogs use a total: Worsley, H. K., and S. J. O'Hara. 2018. "Cross-Species Referential Signalling Events in Domestic Dogs (*Canis familiaris*)." *Animal Cognition* 21: 457–65.

context and tone: Crockford, C., and C. Boesch. 2003. "Context-Specific Calls in Wild Chimpanzees." *Animal Behavior* 66: 115–25.

Chimpanzee gesture types are each used: Kersken et al., "A Gestural Repertoire of 1- to 2-Year-Old Human Children."

only twenty or so meanings: Byrne, R. W., et

al. 2017. "Great Ape Gestures: Intentional Communication with a Rich Set of Innate Signals." *Animal Cognition.* Online.

of those same *gestures:* Kersken et al., "A Gestural Repertoire of 1- to 2-Year-Old Human Children."

overlap 90 percent: Hobaiter, C. Manuscript. "Gestural Communication in the Great Apes: Tracing the Origins of Language." On Bonobo and chimp gestural overlap, see also: Kirsty, E., et al. 2017. "The Gestural Repertoire of the Wild Bonobo (*Pan paniscus*): A Mutually Understood Communication System." *Animal Cognition* 20: 171–77.

If a captive orangutan: Cartmill, E., and R. W. Byrne. 2017. "Orangutans Modify Their Gestural Signaling According to Their Audience's Comprehension." *Current Biology* 17: 1345–48.

They usually persist: Hobaiter, C., et al. 2017. "Wild Chimpanzees' Use of Single and Combined Vocal and Gestural Signals." *Behavioral Ecology and Sociobiology* 71: 96.

Washoe: Hill, J. H. 1978. "Apes and Language." *Annual Review of Anthropology* 7: 89–112. See also: Griebel, U., et al. 2016. "Developmental Plasticity and Language: A Comparative Perspective." *Topics in Cognitive Science* 8: 435–45.

Loulis: Fouts, R., et al. 1989. "The Infant Loulis Learns Signs from Cross-Fostered

Chimpanzees." In *Teaching Sign Language to Chimpanzees,* ed. R. A. Gardner, B. T. Gardner, and T. E. Van Cantfort (Albany: State University of New York Press), pp. 280–92. See also: Caro, T. M., and M. D. Hauser. 1992. "Is There Teaching in Nonhuman Animals?" *Quarterly Review of Biology* 67: 151–74.

"molding": Gardner, R. A., et al. 1989. *Teaching Sign Language to Chimpanzee* (Albany: State University of New York Press), p. 18.

candy bar: Fouts, R., et al., "The Infant Loulis Learns Signs from Cross-Fostered Chimpanzees," p. 286.

Kanzi: Raffaele, P. 2006. "Speaking Bonobo." *Smithsonian.* Online.

"As a consequence": Pepperberg, I. M. 2017. "Animal Language Studies: What Happened?" *Psychonomic Bulletin and Review* 24: 181–85.

Peace, Six

rhinovirus: Scully, Erik J., et al. 2018. "Lethal Respiratory Disease Associated with Human Rhinovirus C in Wild Chimpanzees, Uganda, 2013." *Emerging Infectious Diseases* 24, no. 2: 267.

A baby under five: Stanford, *New Chimpanzee,* pp. 121–22.

"They simply can't make it": Boesch, *Real Chimpanzee,* p. 48.

Newly orphaned siblings: Hobaiter, C. 2014. " 'Adoption' by Maternal Siblings in Wild Chimpanzees." *PLOS One* 9: e103777.

adult males not only adopt: Boesch, *Real Chimpanzee,* p. 49.

"very tragic" and *"like an accusation":* Jarvis, B. 2018. "Baby Orca." *New York Times.* Online.

carried her dead baby for twenty-seven days: Anderson, J. R., et al. 2010. "Pan Thanatology." *Current Biology* 20: R349–51.

Incas believed: Mann, C. C. 2005. *1491: New Revelations of the Americas Before Columbus* (New York: Knopf), pp. 98–99.

When a chimp died in a fall: Teleki, G. 1973. "Group Response to the Accidental Death of a Chimpanzee in Gombe National Park, Tanzania." *Folia Primatologica* 20: 81–94.

"Chimpanzees' awareness of death has been underestimated": Anderson et al., "Pan thanatology."

Ruda, was dying: Reynolds, *Chimpanzees of the Budongo Forest,* pp. 47, 54.

"log dolls": BBC video. *The Young Chimpanzees That Play with Dolls;* http://www.bbc.com/reel/playlist/a-fairer-world?vpid=p03rw3rw.

leaf-bundle "dolls": van Schaik, C. P., et al. 2003. "Orangutan Cultures and the Evolution of Material Culture." *Science* 299: 102–5.

"If I pretend": Ladygina-Kohts, N. N. 1935. *Infant Chimpanzee and Human Child,* ed. F.M.B. de Waal (repr., New York: Oxford University Press, 2002), p. 121.

rats will often open: Underwood, E. 2015. "Rats Forsake Chocolate to Save a Drowning Companion." *Science.* Online.

more likely to sound an alarm: Crockford, C., et al. 2012. "Wild Chimpanzees Inform Ignorant Group Members of Danger." *Current Biology* 22: 142–46.

too interested in a potentially dangerous snake and *Washoe* and *Zoo chimps:* de Waal, *Mama's Last Hug,* pp. 114–19.

rush and attack: Boesch, *Real Chimpanzee,* pp. 50–52.

"This comes much closer": de Waal, *Mama's Last Hug,* p. 215.

a token of one color and *Chimps also pass tools:* Horner, V., et al. 2011. "Spontaneous Prosocial Choice by Chimpanzees." *Proceedings of the National Academy of Sciences* 108: 13847–51. See also: Yamamoto, S., et al. 2012. "Chimpanzees' Flexible Targeted Helping Based on an Understanding of Conspecifics' Goals." *Proceedings of the National Academy of Sciences* 109: 3588–92. And also: Price, M. 2017. "True Altruism Seen in Chimpanzees, Giving Clues to Evolution of Human Cooperation." *Science.* Online.

The bonobo Panbanisha: de Waal, F. B. M., and F. Lanting. 1997. *Bonobo: The Forgotten Ape* (Berkeley: University of California Press).

"If you ask me": de Waal, *Mama's Last Hug,* p. 217.

bacteria: Godfrey-Smith, *Other Minds,* p. 16.

expectations about social norms: Clay, Z., et al. 2016. "Bonobos (*Pan paniscus*) Vocally Protest Against Violations of Social Expectations." *Journal of Comparative Psychology* 130: 44–54.

a social brain *capable:* Byrne and Whiten, *Machiavellian Intelligence.*

Peace, Eight

estrus: Nadler, R. D. "Primate Menstrual Cycle." Primate Info Net; http://pin.primate.wisc.edu/aboutp/anat/menstrual.html.

over 80 percent of their life: Boesch, *Real Chimpanzee,* p. 22.

females interrupted 90 percent: Ibid., pp. 16–17.

easily match the face with the behind: de Waal, F. B. M., and J. J. Pokorny. 2008. "Faces and Behinds: Chimpanzee Sex Perception." *Advanced Science Letters* 1: 99–103.

move human signaling up: Kret, M. E., et al. 2016. "Getting to the Bottom of Face Processing: Species-Specific Inversion Effects for Faces and Behinds in Humans and Chimpanzees (*Pan troglodytes*)." *PLOS One* 11: e0165357. Online.

someone being hassled: Hobaiter, C. Manuscript. "Gestural Communication in the Great Apes: Tracing the Origins of Language."

Meat sharing and *a different thing for males and females:* Stanford, *New Chimpanzee*, pp. 44, 150.

"the consoler in chief": BBC video, *How Chimpanzees Reveal the Roots of Human Behavior.* Online.

Alpha males give and receive more gestures: Hobaiter, C., et al. 2017. "Wild Chimpanzees' Use of Single and Combined Vocal and Gestural Signals." *Behavioral Ecology and Sociobiology* 71: 96.

Peace, Nine

To make war: Boesch, *Real Chimpanzee*, p. 2.

savagely killing them: Reynolds, *Chimpanzees of the Budongo Forest*, p. 124.

mortality rates due to warfare: Boesch, *Real Chimpanzee*, pp. 141, 156.

Peace, Ten

assessing her familiars and any strangers: Kojima S., et al. 2003. "Identification of Vocalizers by Pant Hoots, Pant Grunts and

Screams in a Chimpanzee." *Primates* 44: 225–30. See also: Sliwa, J., et al. 2011. "Spontaneous Voice–Face Identity Matching by Rhesus Monkeys for Familiar Conspecifics and Humans." *Proceedings of the National Academy of Sciences* 108: 1735.

a play face: Stanford, *New Chimpanzee,* p. 119.

Social difficulties are inevitable: Clay, Z., et al, "Bonobos (*Pan paniscus*) Vocally Protest Against Violations of Social Expectations," 44–54.

reflects a sense of community concern: von Rohr, C. R., et al. 2012. "Impartial Third-Party Interventions in Captive Chimpanzees: A Reflection of Community Concern." *PLOS One* 7: e32494.

more aggressive and *more conciliatory:* de Waal, F.B.M. 2005. "A Century of Getting to Know the Chimpanzee." *Nature* 437: 56–59.

"triadic awareness": de Waal, *Mama's Last Hug,* p. 31.

tending his victim's wounds and *"What feeds guilt and shame":* Ibid., pp. 152–54.

Peace, Eleven

a "thinking society": Reynolds, *Chimpanzees of the Budongo Forest,* p. 110.

four genetically differing kinds of chimpanzees: Last, C. 2012. "Are Western Chimpanzees a New Species of *Pan?*" *Scientific American Blogs.* Online.

no such creature: McGrew, *Cultured Chimpanzee,* p. 1.

Benin, Togo, Burkina Faso: Ginn, L., and K.A.I. Nekaris. 2012. "Strong Evidence That West African Chimpanzee Is Extirpated from Burkina Faso." Poster, Primate Society of Great Britain. Online.

Sordid primate pet trade: Gettleman, J. 2017. "Smuggled, Beaten and Drugged: The Illicit Global Ape Trade." *New York Times,* November 4. Online.

"rates of declines": Kühl, H. S., et al. 2019. "Human Impact Erodes Chimpanzee Behavioral Diversity." *Science* 363: 1453–55.

chimpanzee and gorilla numbers: Pilcher, H. 2005. "Peter Walsh: Going Ape." *Nature* 437: 22.

"choices that are bad for our species": Raz, G. 2014. "Louis Leakey: Where Did Human Beings Originate?" TED Radio Hour transcript. Online.

"Considerable effort is urgently needed:" Kühl et al., "Human Impact Erodes Chimpanzee Behavioral Diversity."

various birds: Diaz, S., et al. 2019. "Summary for Policymakers of the Global Assessment Report on Biodiversity and Ecosystem Services of the Intergovernmental Science-Policy Platform on Biodiversity and Ecosystem Services." May 6. Online.

Roughly 80 percent of marine mammals: Valdivia, A., et al. 2019. "Marine Mammals and Sea Turtles Listed Under the U.S. Endangered Species Act Are Recovering." *PLOS One* 14: e0210164.

nine giraffe subspecies: Platt, J. 2009. "Nearly Extinct Giraffe Subspecies Enjoys Conservation Success." *Scientific American*'s *Extinction Countdown* blog. November 13.

roughly 99 percent of their natural lives: Stanford, *New Chimpanzee,* p. 66.

Epilogue

Spix's macaw: Butchart, Stuart H. M., et al. 2018. "Which Bird Species Have Gone Extinct? A Novel Quantitative Classification Approach." *Biological Conservation* 227: 9–18.

SELECTED BIBLIOGRAPHY

Ackerman, J. 2016. *The Genius of Birds.* New York: Penguin.

Beale, Thomas. 1839. *The Natural History of the Sperm Whale . . . To Which Is Added, A Sketch of a South-Sea Whaling Voyage.* Holland Press. Online at Archive.org.

Birkhead, T. 2012. *Bird Sense.* London: Walker.

Boesch, C. 2009. *The Real Chimpanzee.* Cambridge: Cambridge University Press.

Burger, J. 2001. *The Parrot Who Owns Me.* New York: Villard.

Byrne, R., and A. Whiten. 1989. *Machiavellian Intelligence: Social Expertise and the Evolution of Intellect in Monkeys, Apes, and Humans.* New York: Oxford University Press.

Cousteau, J. Y., and F. Dumas. 1953. Reprint, 2004. *The Silent World.* New York: Penguin Random House.

Darwin, C. 1859. *On the Origin of Species by Means of Natural Selection; or, The Preserva-*

tion of Favoured Races in the Struggle for Life. London: John Murray.

Darwin C. 1871. The Descent of Man, and Selection in Relation to Sex. London: John Murray.

de Waal, F. B. M. 2019. Mama's Last Hug. New York: W. W. Norton.

Ellis, R. 2011. The Great Sperm Whale. Lawrence: University Press of Kansas.

Godfrey-Smith, P. 2016. Other Minds. New York: Farrar, Straus and Giroux.

Herzing, Denise L. 2011. Dolphin Diaries. New York: St. Martin's.

Lumsden C. J., and E. O. Wilson. 1981. Genes, Mind and Culture. Cambridge, MA: Harvard University Press.

Mann, Charles C. 2005. 1491: New Revelations of the Americas Before Columbus. New York: Knopf.

Mann, J., R. C. Connor, P. L. Tyack, and H. Whitehead. 2000. Cetacean Societies. Chicago: University of Chicago Press.

Matthiessen, P. 1971. Blue Meridians. New York: Penguin.

McGrew, W. 2004. The Cultured Chimpanzee. Cambridge: Cambridge University Press.

Melville, H. 1851. Reprint, 2003. Moby-Dick. Dover Thrift Editions.

Reiss, D. 2011. The Dolphin in the Mirror. Boston: Houghton Mifflin Harcourt.

Reynolds, V. 2005. The Chimpanzees of the

Budongo Forest. Oxford: Oxford University Press.

Rothenberg, D. 2010. *Thousand Mile Song.* New York: Basic Books.

Stanford, C. 2018. *The New Chimpanzee.* Cambridge, MA: Harvard University Press.

Whitehead, H. 1990. *Voyage to the Whales.* White River Junction, VT: Chelsea Green.

Whitehead, H. 2003. *Sperm Whales: Social Evolution in the Ocean.* Chicago: University of Chicago Press.

Whitehead, H., and L. Rendell. 2015. *The Cultural Lives of Whales and Dolphins.* Chicago: University of Chicago Press.

ACKNOWLEDGMENTS

Praise be to: Shane Gero and all his compatriots of the Dominica Sperm Whale Project. Praise to the exceptional Cat Hobaiter and the generous welcome I received at the Budongo Conservation Field Station in Uganda and, especially, Geoffrey Muhanguzi, Kizza Vincent, Robert Eguma, Monday Gideon, and Pawel Fedurek. Praise be to Donald J. Brightsmith, Gaby Vigo, Ines Duran, Varun Swamy, Kurt Holle, Gabriela Orihuela, Rainforest Expeditions, and the superb staff at Refugio Amazonas and Tambopata Research Center, who were just unreasonably generous. Sam Williams of the Macaw Recovery Network helped me gain unexpected insights. Praise to Denise Herzing and the Wild Dolphin Project; Denise was more accommodating than the weather, and I regret that my time with the dolphins was too brief, though the solar eclipse sunglasses were cool. Likewise my perspective was tweaked by Ben Kilham, Phoebe Kilham, Debbie Kilham, and the mind-altering black bear neighbors of their forests. Barrie Gilbert, whose mind was opened by a grizzly bear, offered thought-provoking conversation and sound sounding, as well as a northbound, five-hundred-mile-long place to work on manuscript revisions. Andrew Rowan was kind enough to inform me about chimpanzee rehabilitation in Guinea. Anyone interested in the creatures in these pages should consider supporting, working with, or visiting the people and places above. Your help would be important.

For crucial support, I am indebted to Susan O'Connor, Roy O'Connor, the Prop Foundation and the Charles Engelhard Foundation, Anne E. and her marine island team, the Gilchrist family and the Wallace Research Foundation, Stony Brook University and its School of Marine and Atmospheric Sciences, the Andrew Sabin Family Foundation, the Kendeda Fund, Ann Hunter-Welborn and David Welborn and family, the momentous Julie Packard, Robert Campbell, Yvon Chouinard, Peter Neumeier, Alfred and Jane Ross, Sven Lindblad, Roslyn and Jerome Meyer, Sunita Chaudry, Sunshine Comes First, Avalon Preserve, and many kindred comrades.

My indefatigable agent Jennifer Weltz is always steadfast, keeping various projects on the rails. My editor Barbara Jones is always ready for a sounding session. And agent and editor emeriti — the stalwart Jean Naggar and the astonishing Jack Macrae — graciously remain only a phone call away. My friends Paul Greenberg and Deborah Milmerstadt generously read and commented on early drafts. Chris Haak provided stimulating intellectual repartee and important academic literature. I literally would not know where I would be without Mayra Mariño, who so gracefully handles all my travel arrangements and so much else.

My wife, Patricia Paladines, displays exceptional patience with my absences and more so by enduring the many inconveniences I cause upon my return. I thank our doggies for keeping the smile in our hearts and their wet tongues on our cheeks. They remind us daily what it looks like to love being alive. We try our best to emulate their example. There is joy and great beauty in it all.

ABOUT THE AUTHOR

Carl Safina's writing about the living world has won a MacArthur "genius" prize and Pew and Guggenheim fellowships; book awards from Lannan, Orion, and the National Academies; and the John Burroughs, James Beard, and George Rabb medals. He earned a PhD in ecology from Rutgers. Safina is the first Endowed Professor for Nature and Humanity at Stony Brook University, where he cochairs the Alan Alda Center for Communicating Science and runs the not-for-profit Safina Center. He hosted the PBS series *Saving the Ocean.* His writing appears in *The New York Times, Time, Audubon,* and on the Web at National Geographic News and Views, *The Huffington Post,* CNN.com, and elsewhere. This is Carl's eighth book. He lives on Long Island, New York, with his wife, Patricia, and their dogs and feathered friends.

The employees of Thorndike Press hope you have enjoyed this Large Print book. All our Thorndike, Wheeler, and Kennebec Large Print titles are designed for easy reading, and all our books are made to last. Other Thorndike Press Large Print books are available at your library, through selected bookstores, or directly from us.

For information about titles, please call:
(800) 223-1244

or visit our website at:
gale.com/thorndike

To share your comments, please write:
Publisher
Thorndike Press
10 Water St., Suite 310
Waterville, ME 04901